AS/A-LEVEL YEAR 1
STUDENT GUIDE

EDEXCEL

Physics

Topics 2 and 3
Mechanics
Electric circuits

Mike Benn

Special thanks to Graham George for his help and advice in the publication of this book.

Philip Allan, an imprint of Hodder Education, an Hachette UK company, Blenheim Court, George Street, Banbury, Oxfordshire OX16 5BH

Orders

Bookpoint Ltd, 130 Milton Park, Abingdon, Oxfordshire OX14 4SB

tel: 01235 827827

fax: 01235 400401

e-mail: education@bookpoint.co.uk

Lines are open 9.00 a.m.–5.00 p.m., Monday to Saturday, with a 24-hour message answering service. You can also order through the Hodder Education website: www.hoddereducation.co.uk

© Mike Benn 2015

ISBN 978-1-4718-4339-6

First printed 2015

Impression number 5 4 3 2 1

Year 2018 2017 2016 2015

All rights reserved; no part of this publication may be reproduced, stored in a retrieval system, or transmitted, in any other form or by any means, electronic, mechanical, photocopying, recording or otherwise without either the prior written permission of Hodder Education or a licence permitting restricted copying in the United Kingdom issued by the Copyright Licensing Agency Ltd, Saffron House, 6–10 Kirby Street, London EC1N 8TS.

This guide has been written specifically to support students preparing for the Edexcel AS and A-level physics examinations. The content has been neither approved nor endorsed by Edexcel and remains the sole responsibility of the author.

Cover photo: alexskopje/Fotolia

Typeset by Integra Software Services Pvt. Ltd, Pondicherry, India

Printed in Italy

Hachette UK's policy is to use papers that are natural, renewable and recyclable products and made from wood grown in sustainable forests. The logging and manufacturing processes are expected to conform to the environmental regulations of the country of origin.

Contents

Getting the most from this book 4
About this book 5

Content Guidance

Mechanics 6
    Rectilinear motion 6
    Scalar and vector quantities 13
    Forces 16
    Momentum 23
    Turning forces 28
    Work, energy and power 29

Electric circuits 34
    Series circuits 34
    Parallel circuits 35
    Electrical energy and power 37
    Ohm's law 38
    Resistivity 39
    Potential dividers 41
    Electromotive force 43
    Factors affecting current in conductors 45

Questions & Answers

AS Test Paper 1 51
A-level Test Paper 1 64

Knowledge check answers 78
Index 79

■ Getting the most from this book

Exam tips
Advice on key points in the text to help you learn and recall content, avoid pitfalls, and polish your exam technique in order to boost your grade.

Knowledge check
Rapid-fire questions throughout the Content Guidance section to check your understanding.

Knowledge check answers
1 Turn to the back of the book for the Knowledge check answers.

Summaries
- Each core topic is rounded off by a bullet-list summary for quick-check reference of what you need to know.

Exam-style questions

Commentary on the questions
Tips on what you need to do to gain full marks, indicated by the icon ⓔ.

Sample student answers
Practise the questions, then look at the student answers that follow.

Commentary on sample student answers
Find out how many marks each answer would be awarded in the exam and then read the comments (preceded by the icon ⓔ) following each student answer.

■ About this book

This guide is one of a series covering the Edexcel specification for AS and A-level physics. It offers advice for the effective study of Topics 2 and 3 (Mechanics and Electric circuits). Its aim is to help you *understand* the physics — it is not intended as a shopping list, enabling you to cram for the examination. The guide has two sections:

- The **Content Guidance** is not intended to be a detailed textbook. It offers guidance on the main areas of the content of the topics, with an emphasis on worked examples. These examples illustrate the types of question that you are likely to come across in the examinations.
- The **Questions & Answers** section comprises two sample tests — one with the structure and style of the AS paper 1 examination, the other with the structure of an A-level paper 1. Both tests are restricted to the content of this guide. Answers are provided and, in some cases, distinction is made between responses that might have been given by an A-grade student and those of a typical grade-C student. Common errors made by students are also highlighted so that you, hopefully, do not make the same mistakes.

The purpose of this book is to help you with the AS paper 1 and A-level paper 1 and synoptic paper 3 examinations, but don't forget that what you are doing is learning physics. The development of an understanding of physics can only evolve with experience, which means time spent thinking about physics, working with it and solving problems. This book provides you with the platform to do this.

If you try all the worked examples and the tests before looking at the answers, you will begin to think for yourself as well as develop the necessary technique for answering examination questions. In addition, you will need to learn the basic formulae, definitions and experiments. Thus prepared, you will be able to approach the examination with confidence.

The specification states the physics that will be examined in the AS and A-level examinations and describes the format of those tests. This is not necessarily the same as what teachers might choose to teach (or what you might choose to learn).

The specification can be obtained from Edexcel, either as a printed document or from the web at www.edexcel.com.

Content Guidance

Mechanics

Rectilinear motion

This section covers the motion of objects in a given direction when they are either moving with a constant speed in that direction (i.e. constant velocity), or accelerating at a constant rate.

Equations of motion (*suvat* equations)

The relevant quantities needed to represent the motion of an object in one dimension are:

- displacement, s — m
- initial velocity, u — $m\,s^{-1}$
- final velocity, v — $m\,s^{-1}$
- acceleration, a — $m\,s^{-2}$
- time, t — s

Using these symbols the definitions become:

$$\text{average velocity} = \frac{u+v}{2} = \frac{s}{t}$$

$$\text{acceleration} = \frac{\text{change in velocity}}{\text{time}} = \frac{v-u}{t}$$

Combining the above definitions leads to a set of equations where, if any three of the quantities is known, the other two may be calculated. These are sometimes referred to as the '*suvat*' equations and they are reproduced at the end of the examination in the formulae section.

- $v = u + at$
- $s = ut + \frac{1}{2}at^2$
- $v^2 = u^2 + 2as$

Average velocity is displacement divided by time.

Acceleration is the rate of change in velocity.

Exam tip

These equations can only be used for bodies moving in a **straight line** with **uniform** acceleration.

Worked example

An athlete starts from rest and accelerates uniformly for 4.0 s when she has reached a velocity of $8.0\,m\,s^{-1}$. Calculate:

a her acceleration

b the distance she has travelled during this time

Answer

a Use $v = u + at$
$8.0\,m\,s^{-1} = 0\,m\,s^{-1} + a \times 4.0\,s$
$a = 2.0\,m\,s^{-2}$

b Use $s = ut + \frac{1}{2}at^2$

$$s = 0\,\text{m}\,\text{s}^{-1} \times 4.0\,\text{s} + \frac{1}{2} \times 2.0\,\text{m}\,\text{s}^{-2} \times (4.0\,\text{s})^2$$

$$= 16\,\text{m}$$

> **Exam tip**
>
> When using the 'suvat' equations it is useful to write down all the quantities with the given values, in this case:
>
> $u = 0\,\text{m}\,\text{s}^{-1}$, $v = 8.0\,\text{m}\,\text{s}^{-1}$, $t = 4.0\,\text{s}$; a and s are to be found.
>
> Therefore use the equation $v = u + at$ to find a and then use $v^2 = u^2 + 2as$ or $s = ut + \frac{1}{2}at^2$ to find s.

Acceleration due to gravity

In the late sixteenth century, Galileo Galilei performed his famous experiment, dropping a variety of different sized balls (of different mass) from the Leaning Tower of Pisa. The balls, when released simultaneously, hit the ground at the same time, showing that the acceleration due to gravity is independent of the mass of free-falling objects. Later experiments showed that a coin and a feather in an evacuated glass tube fell at the same rate, and the *Apollo 15* astronaut David Scott performed a similar test on the Moon using a hammer and a feather (Figure 1).

The value of the acceleration due to gravity has been accurately measured on the Earth's surface and, although there are small variations around the globe, the accepted value is usually given as $9.81\,\text{m}\,\text{s}^{-2}$. For the purposes of this guide all calculations use $9.8\,\text{m}\,\text{s}^{-2}$ and if estimates are needed $10\,\text{m}\,\text{s}^{-2}$ is adequate. This acceleration is usually referred to as 'little g'.

> **Knowledge check 1**
>
> State the difference between instantaneous and average velocity.

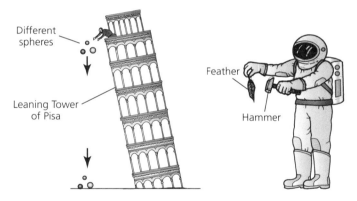

Figure 1 Two classic experiments showing that acceleration due to gravity is independent of the mass of free-falling objects

The value of g can be found by timing an object falling freely from rest through a measured distance, and then using the appropriate equation of motion. In practice, because the distances, and hence the times, are usually relatively small, sensitive timing devices are required. A simple arrangement is shown in Figure 2, where a switch de-energises an electromagnet and simultaneously starts an electronic timer. A steel sphere is released from the magnet and falls onto a 'trap-door switch' that opens to switch off the timer.

The equation $s = ut + \frac{1}{2}at^2$ is applied with s representing the distance fallen by the sphere. As the ball falls from rest, u equals zero and so the acceleration is found by rearranging the equation to give:

$$g = \frac{2s}{t^2}$$

Content Guidance

There are several variations of this experiment. Using light gates with a computer interface, strobe photography and video-frame analysis are examples, but in all cases g is calculated from the time taken for a free-falling object to move through a measured distance.

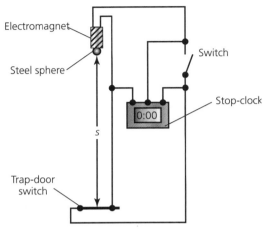

Figure 2 Simple apparatus for measuring g

> **Core practical 1**
>
> **Acceleration of a freely falling object**
> Core practical 1 requires you to determine the acceleration of free fall of a small object by measuring the time for the object to fall from a range of heights. Full details of how the measurements are taken, the precautions taken and how the acceleration is found using a graphical method may be required for the examinations.

Worked example

A stone is dropped into a well. The splash as it enters the water is timed as 1.8 s after release. Calculate:

a the depth of the well

b the velocity of the stone when it strikes the surface of the water

Answer

$u = 0\,\text{m s}^{-1}$ $a = g = 9.8\,\text{m s}^{-2}$ $t = 1.8\,\text{s}$

s (depth) and v are to be found.

a $s = ut + \tfrac{1}{2}at^2 = 0 + \tfrac{1}{2} \times 9.8\,\text{m s}^{-2} \times (1.8\,\text{s})^2 = 16\,\text{m}$

b $v = u + at = 0\,\text{m s}^{-1} + 9.8\,\text{m s}^{-2} \times 1.8\,\text{s} = 18\,\text{m s}^{-1}$

Exam tip

For all free-falling objects the acceleration will be $9.8\,\text{m s}^{-2}$ assuming that the effects of air resistance are negligible.

Mechanics

Horizontal and vertical motion

When a skydiver jumps from a moving aeroplane, he will start to accelerate downwards but will also be moving horizontally at the speed of the aircraft. To explain the trajectory of the diver it is possible to consider the vertical and horizontal motions independently (Figure 3).

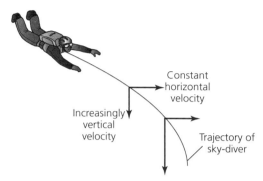

Figure 3

If air resistance is ignored, the skydiver will accelerate downwards with a velocity of $9.8\,\text{m s}^{-2}$ while at the same time continuing to move horizontally at the speed at which he left the plane. Figure 3 illustrates the subsequent parabolic motion. It is possible to calculate the horizontal distance covered by a falling object by applying the equations of motion in the vertical plane to find the time in free fall, and calculating the horizontal distance covered in this time when travelling at constant speed.

Worked example

A sea-eagle, carrying a fish in its talons, is flying horizontally with a velocity of $5.0\,\text{m s}^{-1}$ at a height of $12\,\text{m}$ above the sea. The fish wriggles free and falls back into the water. Calculate:

a the time taken for the fish to reach the sea
b the horizontal displacement of the fish during this time

Answer

a For the vertical motion: $u = 0\,\text{m s}^{-1}$ $\quad a = 9.8\,\text{m s}^{-2} \quad s = 12\,\text{m}$

$s = ut + \tfrac{1}{2}at^2 \quad 12\,\text{m} = 0 + \tfrac{1}{2} \times 9.8\,\text{m s}^{-2} \times t^2 \quad t = 1.6\,\text{s}$

b For the horizontal motion: $s = ut = 5.0\,\text{m s}^{-1} \times 1.6\,\text{s} = 8.0\,\text{m}$

(or $7.8\,\text{m}$ if data are kept in your calculator)

Displacement–time and velocity–time graphs

It is often useful to represent the motion of objects by plotting graphs of their displacement or velocities against time. In addition to illustrating the nature of the motion it is possible to calculate the values of displacement, velocity and acceleration from the graphs.

Knowledge check 2

Why does a ball projected horizontally from a height, h, take the same time to reach the floor as a similar ball dropped vertically from the same height?

Exam tip

The independence of the vertical and horizontal motion of bodies in free fall is an important concept. It allows you to use the equations of motion separately for each direction.

Content Guidance

Displacement–time graphs

Figure 4 represents the displacement–time graph for an object moving at constant velocity.

$$\text{velocity} = \frac{\text{displacement}}{\text{time}} = \frac{\Delta s}{\Delta t} = \text{gradient of the line}$$

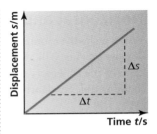

Figure 4 Graph for constant velocity

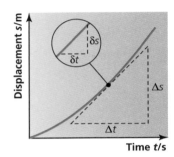

Figure 5 Instantaneous velocity for an accelerating object

In Figure 5 the gradient of the line is increasing with time. This means that the velocity is increasing and so the object is accelerating. The value of the velocity at any instant can be found by measuring the gradient of a tangent drawn at that instant.

$$\text{instantaneous velocity} = \frac{\delta s}{\delta t} = \text{gradient of tangent} = \frac{\Delta s}{\Delta t}$$

Velocity–time graphs

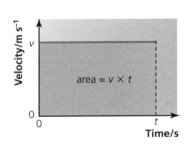

Figure 6

The graphs in Figure 6 represent an object moving at constant speed and another with uniform acceleration.

$$\text{acceleration} = \frac{\text{change in velocity}}{\text{time}} = \text{gradient of the line}$$

The first line has zero gradient, so the velocity is constant, i.e. the acceleration is zero.

The gradient of the second line is constant, so the line represents uniform acceleration. A curve with increasing gradient, like Figure 5, but for a velocity–time graph will indicate an acceleration that is increasing.

Exam tip

When analysing displacement–time and velocity–time graphs always observe how the gradient is changing to visualise the change in velocity.

In the section on the equations for uniform motion we saw that:

displacement = average velocity × time = $\frac{u+v}{2} \times t$

The graphs show that the area under the line of each graph represents the displacement of the object during that period. This is also true for non-uniform motion, when the displacement can be found by estimating the area under the curve.

Acceleration–time graphs

In most cases of rectilinear motion studied at AS and A-level the acceleration of objects is uniform. There are, however, a few instances when the acceleration changes and an acceleration-time graph would be useful for describing the motion of an object.

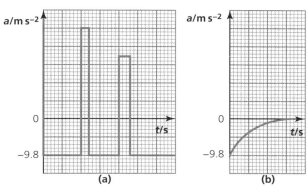

Figure 7

Figure 7(a) represents the motion of a bouncing ball. Between bounces the ball is in free fall and is accelerating downwards at (−) $9.8\,\mathrm{m\,s^{-2}}$ (even when it is moving upwards, the acceleration due to gravity continues). The velocity changes rapidly from downward to upward when the ball is in contact with the floor. This change from a downward velocity to an upward velocity in a short time results in a large (upward) acceleration.

Figure 7(b) represents the motion of an object, starting from rest, falling through a viscous medium. As the object falls its velocity increases. This produces an increasing upward resistive force and so reduces the resultant downward force and hence the acceleration of the object. Eventually the resistive forces equal the weight of the object and it moves with a constant velocity (terminal velocity). This will be looked at later in the fluids section.

It should be emphasised that when the acceleration is zero, the object will be moving at a constant velocity.

Knowledge check 5

Sketch acceleration-time graphs for (a) a steel ball in free-fall close to the Earth's surface, and (b) a similar ball moving at constant velocity.

Knowledge check 3

What quantities are represented by the gradients of displacement–time graphs and velocity–time graphs?

Knowledge check 4

How is the displacement of a moving body determined from a velocity–time graph?

Content Guidance

Worked examples

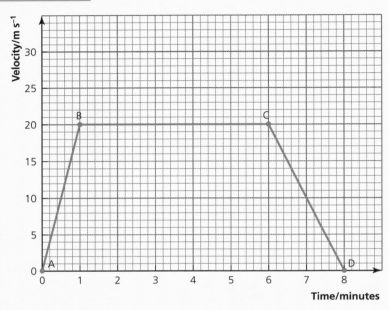

Figure 8

> **Exam tip**
> For uniform motion (as shown in this question) the displacement can be found from the areas of the triangles and rectangle, or directly using the trapezium rule.

> **Exam tip**
> Remember to convert times into seconds in questions like this.

The velocity–time graph in Figure 8 represents the motion of a train as it travels from station A to station D. Describe the changes in the motion of the train, and calculate:

a the acceleration from A to B
b the acceleration from C to D
c the total displacement from A to D

Answer

The train accelerates uniformly from A to B, continues to travel at $20\,\mathrm{m\,s^{-1}}$ until it reaches C, and then decelerates, uniformly, to D.

a acceleration = gradient = $\dfrac{20\,\mathrm{m\,s^{-1}}}{60\,\mathrm{s}} = 0.33\,\mathrm{m\,s^{-2}}$

b acceleration = gradient = $\dfrac{-20\,\mathrm{m\,s^{-1}}}{120\,\mathrm{s}} = -0.17\,\mathrm{m\,s^{-2}}$

(The negative sign indicates that the acceleration is in the opposite direction from the velocity — i.e. it is decelerating.)

c displacement = area under the graph

$= (\tfrac{1}{2} \times 20\,\mathrm{m\,s^{-1}} \times 60\,\mathrm{s}) + (20\,\mathrm{m\,s^{-1}} \times 300\,\mathrm{s}) + (\tfrac{1}{2} \times 20\,\mathrm{m\,s^{-1}} \times 120\,\mathrm{s})$

$= 7800\,\mathrm{m}$

Mechanics

> **Summary**
>
> After studying this topic, you should be able to:
> - use the 'suvat' equations to determine values of initial and final velocities, acceleration, displacement and time, for bodies with uniform motion
> - use displacement–time and velocity–time graphs to describe the motion of an object
> - determine the velocity of a body from the gradient of a displacement–time graph
> - determine the acceleration of a body from the gradient of a velocity–time graph, and its displacement using the area under the graph
> - describe the motion of an object from an acceleration-time graph

Scalar and vector quantities

The difference between **scalar** and **vector quantities** can be explained using Figure 9.

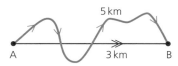

Figure 9

If you walk from A to B using the winding road, you will have travelled a distance of 5 km in no specific direction. This *distance* is therefore a scalar quantity.

On reaching B you will have been displaced from your starting position 3 km in the direction AB. This *displacement* is a vector quantity as it includes not only the size but also the direction.

If the journey took half an hour your average *speed* is the distance divided by the time: 5 km/0.5 h = 10 km h^{-1}. As there is no implied direction it is a scalar quantity.

The average *velocity* is the displacement divided by the time: 3 km/0.5 h = 6 km h^{-1} in the direction AB, and therefore a vector quantity.

> **Scalar** quantities have magnitude (size) only. Examples of scalars include mass, the number of particles of a substance (moles), distance, speed and time.
>
> **Vector** quantities have both magnitude and direction. Examples of vectors include displacement, velocity, acceleration, fields and forces.

> **Worked example**
>
> A ball is thrown vertically upwards with an initial velocity of 8.0 m s^{-1}. Calculate:
> a the time taken for the ball to reach its maximum height
> b the total upward displacement of the ball
>
> **Answer**
>
> Taking upward vectors as positive:
>
> $u = +8.0 \text{ m s}^{-1}$
>
> $v = 0 \text{ m s}^{-1}$ (the ball stops momentarily at the top of its flight)
>
> $a = g = -9.8 \text{ m s}^{-2}$ (negative as the acceleration is in a downward direction)
>
> a $v = u + at$
> $0 \text{ m s}^{-1} = +8.0 \text{ m s}^{-1} - 9.8 \text{ m s}^{-2} \times t \Rightarrow t = 0.82 \text{ s}$
> b $v^2 = u^2 + 2as$
> $(0 \text{ m s}^{-1})^2 = (+8.0 \text{ m s}^{-1})^2 - 2 \times 9.8 \text{ m s}^{-2} \times s \Rightarrow s = 3.3 \text{ m}$

> **Exam tip**
>
> It is important that the *direction* of a vector quantity is stated or clearly indicated on a diagram.

> **Knowledge check 6**
>
> Name two basic scalar quantities and one derived scalar quantity.
>
> Name one basic vector quantity and two derived vector quantities.

Mechanics • Electric circuits 13

Content Guidance

Scalar and vector addition

- Scalar quantities are added using normal arithmetic — for example, if you have 5 kg of potatoes in a basket and you add a further 2 kg of carrots, the total mass of the vegetables is 7 kg.
- To add vector quantities a diagram must be drawn showing both magnitude and direction of the quantities.

Exam tip

s, u, v, and a are all vector quantities, and time t is the only scalar. It is therefore necessary to indicate the direction of the vectors by allocating a positive value (e.g. left to right or upward) and a negative value for quantities acting in the opposite direction.

Worked example

An athlete runs due east for 4 kilometres and then due south for a further 3 kilometres. The run takes a total time of 20 minutes. Calculate:
a the distance travelled by the athlete
b her displacement from start to finish
c her average speed
d her average velocity

Answer

a distance = 4 km + 3 km = 7 km

b The displacement can be found either by scale drawing, representing each leg of the run by a line where 1 cm, say, is equivalent to 1 km, and then measuring the resultant displacement; or, as the vector diagram (Figure 10) includes a right-angle triangle, Pythagoras's theorem and basic trigonometry can be used.

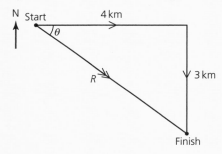

Figure 10

By measurement: $R = 5.0$ cm $= 5$ km; $\theta = 37°$

By calculation: $R^2 = (4\,\text{km})^2 + (3\,\text{km})^2$; $R = 5$ km; $\tan \theta = \frac{3}{4} \Rightarrow \theta = 37°$

The resultant displacement is 5 km in the direction E 37° S

c average speed = 7×10^3 m/1200 s = 5.8 m s^{-1}
d average velocity = 5×10^3 m/1200 s = 4.2 m s^{-1} in the direction E 37° S

Exam tip

Vector diagrams are usually right angle triangles. You should always use a ruler and clearly show the direction of the vectors using arrows.

Resolution of vectors

- A single vector can have the same effect as two other vectors that, when added together, would result in an identical vector.
- It is often convenient to represent a single vector as a pair of **component** vectors at right angles to each other (Figure 11).

V is equivalent to the components V_x and V_y. Using trigonometry:
- $\cos \theta = V_x/V \rightarrow V_x = V \cos \theta$
- $\sin \theta = V_y/V \rightarrow V_y = V \sin \theta$

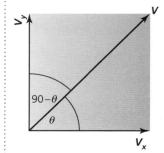

Figure 11

Mechanics

Worked example

A gardener pulls a roller with a force of 200 N. The angle between the ground and the handle is 30°. Calculate the horizontal and vertical components of the force.

Answer

See Figure 12.

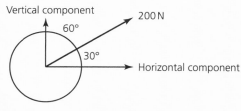

Figure 12

- Horizontal component = 200 N cos 30 = 173 N
- Vertical component = 200 N cos 60 (= 200 N sin 30) = 100 N

The roller behaves as if it is being pulled horizontally with a force of 173 N and being lifted up with a force of 100 N.

Exam tip

You may find it easier to use only the cosine of the angle between the vector and its component. For the example shown, this will give $V_y = V \cos(90 - \theta)$ as an alternative to $V \sin \theta$.

The resolution of a vector into a pair of components at right angles is a useful tool for many applications in physics. You will come across several applications later in this guide relating to the resolution of forces, and the A-level course requires the use of components of electric and magnetic fields.

You have seen that for bodies moving freely under gravity the horizontal and vertical motions can be treated independently. In the study of projectiles it is possible to gain information about the trajectories by taking horizontal and vertical components of the velocities (Figure 13).

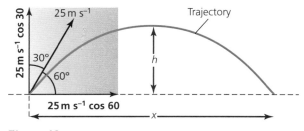

Figure 13

The initial velocity has components $u \cos \theta$ in the horizontal plane, and $u \cos(90 - \theta)$ (or $u \sin \theta$) in the vertical plane.

The vertical motion is always accelerating downwards at $9.8 \, \text{m s}^{-2}$ whereas the horizontal component of the velocity remains the same throughout the flight of the projectile.

Applying the equations of motion to the vertical components you can calculate the maximum height reached and the time of the flight, and then find the range using the horizontal component of the velocity.

Knowledge check 7

Why does the horizontal component of the velocity of a projectile remain the same throughout the motion?

Content Guidance

> **Worked example**
>
> A cricket ball is struck so that the ball leaves the bat at 60° to the ground travelling at $25\,\text{m s}^{-1}$. Calculate:
>
> a the maximum height reached by the ball (assume it starts at ground level)
> b the total time the ball is in the air
> c the horizontal distance from the bat when the ball first hits the ground
>
> **Answer**
>
> In the vertical plane: In the horizontal plane:
> $u = 25\,\text{m s}^{-1}\cos 30 = 22\,\text{m s}^{-1}$ $u = 25\,\text{m s}^{-1}\cos 60 = 12.5\,\text{m s}^{-1}$
> $v = 0\,\text{m s}^{-1}$ (at the highest point, $s = h$)
> $a = g = -9.8\,\text{m s}^{-2}$
>
> a $v^2 = u^2 + 2as$ $(0\,\text{m s}^{-1})^2 = (22\,\text{m s}^{-1})^2 - 2 \times 9.8\,\text{m s}^{-2} \times h \Rightarrow h = 25\,\text{m}$
> b $v = u + at$ $0\,\text{m s}^{-1} = 22\,\text{m s}^{-1} - 9.8\,\text{m s}^{-2} \times t \Rightarrow t = 2.2\,\text{s}$
>   This is the time to the top of the trajectory, so the total time in the air will be $4.4\,\text{s}$.
> c The ball travels at a constant horizontal velocity of $12.5\,\text{m s}^{-1}$:
>   horizontal displacement, $x = 12.5\,\text{m s}^{-1} \times 4.4\,\text{s} = 55\,\text{m}$

Exam tip

In all projectile calculations the horizontal and vertical motions are independent of each other.

> **Summary**
>
> After studying this topic, you should be able to:
> - differentiate between scalar and vector quantities
> - identify scalars and vectors from a list of basic and derived quantities
> - draw vector diagrams to represent the addition of two or more vectors
> - determine the resultant of two vectors using scale drawing or trigonometric calculations
> - resolve a vector into two components at right angles to each other

Forces

Forces push, pull, squeeze or stretch. They fall into two categories: distant and contact forces.

- Forces acting over a distance: gravitational (e.g. between the Sun and the planets) and electromagnetic (between static and moving charges, and between magnetic poles). Nuclear forces are not considered at AS, but will be discussed in the student guide covering Topics 9–13 in this series.
- Contact forces: reaction forces between your shoe and the floor, friction, viscosity and air resistance.

Tension forces, such as the forces tending to restore the initial length of a stretched rubber band, are caused by short-range attractive forces between displaced molecules. These are electromagnetic in nature and are considered more fully under materials in the student guide covering Topics 4 and 5 in this series.

Mechanics

Newton's first law of motion

An object in outer space, if not subjected to any forces, will stay at rest or continue to move at constant velocity. Its motion will also be unaffected if any forces acting on the object cancel out. This principle was first suggested by Isaac Newton, and is now accepted as **Newton's first law of motion**.

Forces are vector quantities. You have seen that the addition of forces requires the direction of the forces to be considered as well as their magnitudes. The total sum of a number of forces, acting at a point, is termed the *resultant* force. When the resultant force acting on a body is zero, the body is said to be in *equilibrium*.

In Figure 14, for P to be in equilibrium the vector sum of F_1, F_2 and F_3 must be zero. For this to be true, the sum of the horizontal components must be zero, and also those in the vertical plane. (In fact a body will be in equilibrium if the components of the forces in any two planes add up to zero, but it is most convenient to use components in planes at right angles.)

In the horizontal plane: $(F_2 \cos \theta_2 + F_3 \cos \theta_3) - F_1 = 0$

In the vertical plane: $F_2 \cos (90 - \theta_2) - F_3 \cos (90 - \theta_3) = 0$

> **Newton's first law of motion** states that a body will remain at rest or move with uniform velocity unless acted upon by a resultant external force.

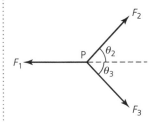

Figure 14 Forces acting on a point in equilibrium

Worked example

A sledge is being pulled across the snow at constant velocity by two husky dogs. One dog pulls with a force of 100 N at an angle of 20° to the direction of travel and the second animal pulls on its harness at an angle of 35° to that of the other dog (Figure 15).

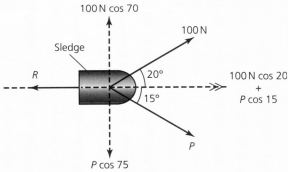

Figure 15

Calculate:
- **a** the force applied to the sledge by the second dog
- **b** the resistive force of the snow on the sleigh

Answer

As the sledge is moving at constant velocity it is in equilibrium, i.e. the resultant of the forces of the dogs and the resistive force of the snow must be zero.

- **a** In the plane perpendicular to the motion: $100\,\text{N} \cos 70 - P \cos 75 = 0 \Rightarrow P = 132\,\text{N}$
- **b** In the direction of motion: $(100\,\text{N} \cos 20 + 132\,\text{N} \cos 15) - R = 0 \Rightarrow R = 221\,\text{N}$

> **Knowledge check 8**
>
> What is meant by the term *equilibrium*?

Content Guidance

Free-body force diagrams

In the above example we considered only the horizontal forces acting on the sledge. In fact there are several other forces acting in the system; the weight of the sledge, the reaction of the ground on the sledge and dogs etc. When considering the equilibrium of a single body we need to isolate all the forces acting on it using a free-body force diagram.

Consider the trapeze artist shown in Figure 16. She 'feels' only three forces: her weight pulling down, the horizontal pull of her assistant and the trapeze rope through her hands.

> **Exam tip**
>
> All the forces on a free-body diagram should act at a single point, be represented by straight lines (use a ruler!) and their directions indicated using arrowheads.

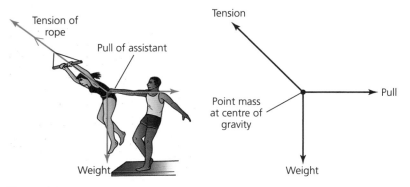

Figure 16

In order to use a free-body force diagram to find components and make calculations, the lines of action of the forces must be accurately drawn. It is clear that the pull of the rope is through the hands and arms and the assistant is holding the belt, but where is the weight acting? The diagram shows all the forces acting at a single point. For all extended bodies there is a point through which the weight always acts. This is known as the **centre of gravity** of the body. For free-body force diagrams it is recommended that the body is represented by a point at its centre of gravity, and all forces are drawn through this point (Figure 17).

> The **centre of gravity** of a body is the point at which the total weight of the body can be said to act.

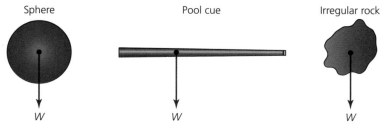

Figure 17 Examples of centre of gravity

The position of the centre of gravity for three different objects is shown above. For regular objects like the sphere the centre of gravity is at the geometric centre, but its position in irregular bodies depends on the shape and the mass distribution. The centre of gravity will always be directly above the balance point (or vertically below the point of suspension) of the body. If you balance a pool cue on your finger and close your eyes, it would feel the same as a single point mass placed on your finger.

> **Knowledge check 9**
>
> How would you estimate the position of the centre of gravity of a laboratory/retort stand?

Mechanics

Newton's second law of motion

Newton's first law tells us that a body will remain at rest or continue to move with constant velocity if no resultant forces act on it. So what will happen if a resultant force is applied? There will be a change in its motion, its velocity will change, which means that the body will accelerate. For *fixed masses* the relationship between the resultant force and the acceleration is given by the expression:

$$F = ma$$

This is an expression of **Newton's second law of motion** applied to bodies of fixed mass, and assumes the force to be in newtons, the mass in kilograms and the acceleration to be in metres per second squared. The definition given is sufficient for most instances in the AS and A-level examinations. A more general definition of the law is given in the student guide covering Topics 6–8 in this series, but only forces acting on fixed masses will be examined here.

Newton's second law of motion states that, for a body of fixed mass, the acceleration is directly proportional to the *resultant* (net) force applied to the body.

Exam tip
Always calculate the resultant force before using the equation $F = ma$.

Worked example

An experiment to investigate the relationship between the acceleration of a fixed mass uses a glider on an air track and two optical timing gates, as shown in Figure 18.

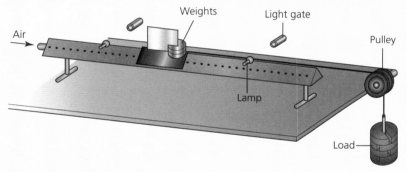

Figure 18 Investigating the acceleration of an object

a The light gates are connected to a computer that measures the times taken for the interrupter card to pass through each gate, and the time for the glider to travel from the first gate to the second one. Describe how you would perform the experiment, explaining how values of force and acceleration are obtained.

b A typical set of readings is given in the table below. F represents the resultant force applied to the mass, t_1 and t_2 are the times for the card to cut through the gates, and t is the time taken by the glider to travel between the gates. Copy the table and complete the columns for the initial and final velocities, u and v, and for the acceleration, a. The length of the interrupter card is 20 cm.

F/N	t_1/s	t_2/s	t/s	u/m s^{-1}	v/m s^{-1}	a/m s^{-2}
0.10	1.25	0.48	2.00			
0.20	0.91	0.34	1.48			
0.30	0.74	0.28	1.18			
0.40	0.63	0.24	1.00			
0.50	0.57	0.23	0.84			

Content Guidance

c Draw a graph of $a/\text{m s}^{-2}$ against F/N. Use your graph to determine the accelerated mass, m.

d Calculate the acceleration of a car of mass 1500 kg if the driving force on the car is 1200 N and the resistive forces (friction, drag) total 450 N.

Answer

a The resultant force acting on the system (the glider plus all the masses) is the downward weight of the load. To increase this force a mass is taken off the glider and added to the load. Thus the total mass of the system stays the same. The velocities of the glider at each gate, u and v, are found by dividing the length of the card by the times to cross the gates, and the acceleration is calculated using the equation $v = u + at$.

b

F/N	t_1/s	t_2/s	t/s	u/m s^{-1}	v/m s^{-1}	a/m s^{-2}
0.10	1.25	0.48	2.00	0.16	0.42	0.13
0.20	0.91	0.34	1.48	0.22	0.59	0.25
0.30	0.74	0.28	1.18	0.27	0.71	0.37
0.40	0.63	0.24	1.00	0.32	0.83	0.51
0.50	0.57	0.23	0.84	0.35	0.87	0.62

c See Figure 19.

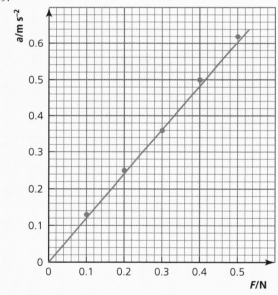

Figure 19

The gradient of the graph = a/F and therefore:

$$m = \frac{F}{a} = \frac{1}{\text{gradient}} = \frac{0.50\,\text{N}}{0.62\,\text{m s}^{-2}} = 0.81\,\text{kg}$$

d Resultant force acting on the car, $\Sigma F = 1200\,\text{N} - 450\,\text{N} = 750\,\text{N}$

$$\Sigma F = ma \Rightarrow a = \frac{750\,\text{N}}{1500\,\text{kg}} = 0.5\,\text{m s}^{-2}$$

Mechanics

Unit of force: the newton

Up to this point in this guide we have just accepted the use of the newton (N) as the SI unit of force. Newton's second law equation for a fixed mass requires the mass to be in kilograms, the acceleration in metres per second squared and the resultant force in **newtons**. It follows from the equation that a resultant force of 1 newton acting on a mass of 1 kilogram will accelerate the mass at a rate of 1 metre per second squared.

> 1 **newton** (N) is that resultant force which, when acting on a mass of 1 kilogram, produces an acceleration of 1 metre per second squared.

Gravitational field strength

The section relating to free-falling bodies using the ideas of Galileo and subsequent experimental evidence showed that objects accelerate at $9.8\,\mathrm{m\,s^{-2}}$ when released close to the Earth's surface (ignoring air resistance). Applying Newton's second law equation to a mass, m, close to the surface:

$$\text{acceleration due to gravity, } g = \frac{F}{m} = 9.8\,\mathrm{m\,s^{-2}}$$

The concept of a field in physics relates to a region where a force is experienced. A gravitational field is therefore a region where a gravitational force is felt, i.e. a mass placed in the field will experience a force. The strength of a gravitational field will determine the size of the force acting upon a given mass. Hence the above expression for the acceleration due to gravity also acts as a measure of **gravitational field strength**.

$$\text{gravitational field strength, } g = \frac{F}{m} = 9.8\,\mathrm{N\,kg^{-1}}$$

> **Gravitational field strength** is the force per unit mass acting on a body in a gravitational field.

The gravitational field strength on the surface of the Earth varies a little. It is slightly greater in the polar regions than on the Equator and can be affected by high-density metallic ores etc. The value of g also decreases with height above the surface such that a rocket reaching 50 000 km will experience little gravitational force from the Earth. The gravitational field strength on the Moon is about $1.6\,\mathrm{N\,kg^{-1}}$, one-sixth of that on the Earth.

Weight is the force acting upon a mass in a gravitational field. Rearranging the expression for the field strength:

$$W = mg$$

So a mass of 1 kg has a weight of 9.8 N on Earth and a weight of 1.6 N on the Moon.

> **Knowledge check 10**
>
> Explain why an object on the surface of the Moon has a smaller weight than if it were on Earth, even though its mass will be the same.

Newton's third law of motion

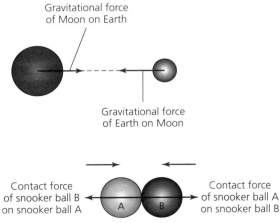

Figure 20 Newton's third law pairs

Content Guidance

The examples shown in Figure 20 illustrate **Newton's third law**. Each shows a pair of forces acting on two different bodies. These pairs must always:

- act on two separate bodies
- be of the same type
- act along the same line
- be equal in magnitude
- act in opposite directions

Newton's third law of motion states that if a body A exerts a force on a body B, then body B will exert an equal and opposite force on body A.

Sometimes there can be more than one pair of forces acting on two bodies. Consider a woman standing on the Earth (Figure 21):

Exam tip

It must be stressed that this law cannot be applied to single bodies, and you should always state which bodies the forces act upon and the direction of the forces.

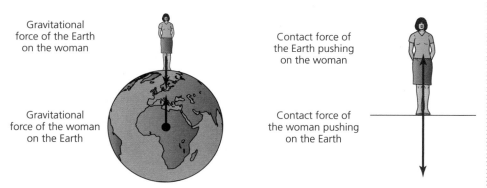

Figure 21 Newton's third law pair forces acting between the Earth and a woman

The diagram shows the two Newton's third law pairs acting on the woman and the Earth:

- gravitational force of the Earth on the woman (down); gravitational force of the woman on the Earth (up)
- contact (elastic) force of the woman pushing down on the Earth; contact force of the Earth pushing up on the woman

If the woman alone is considered, it can be seen that she is in equilibrium due to the action of the two different types of force (Figure 22).

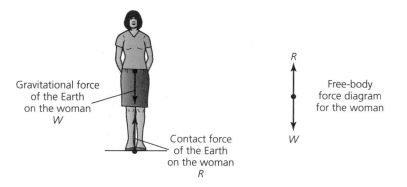

Figure 22 Free-body force diagram for the woman

The resultant of the upward contact force and the downward gravitational force (her weight) is zero, and so she will remain at rest according to Newton's first law.

Exam tip

It is a common error to regard two equal forces acting on a single body as a third law pair, so you must remember that two separate bodies are needed for Newton's third law.

Mechanics

Worked example

A fisherman checks the weight of his catch using a newton-meter. Two of the forces acting are shown in Figure 23.

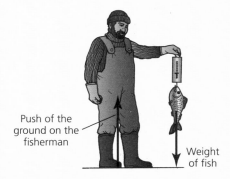

Figure 23

a For each force there is a 'Newton's third law pair force'. In each case state:
  i   the body that the Newton's third law pair force acts upon
  ii the type of force
  iii the direction of the Newton's third law pair force
b Draw a free-body diagram for the fish, labelling the forces acting upon it.

Answer

a

Force	i	ii	iii
Weight of the fish	(Pull of the fish on) the Earth	Gravitational	Upwards
Push of the ground on the fisherman	(Push of the fisherman on) the ground	Contact (elastic) force	Downwards

b See Figure 24.

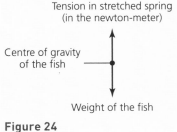

Figure 24

Momentum

Any mass that is moving has **momentum**. The momentum of a body depends on the values of its mass and velocity. For example, a charging elephant has a much bigger momentum than a scurrying mouse.

Knowledge check 11

State Newton's laws of motion.

momentum = mass × velocity
$p = mv$
units: $kg\,m\,s^{-1}$

Content Guidance

Momentum is a **vector** quantity, with both magnitude and direction.

Worked example

a Calculate the (magnitude of the) momentum of:
   i a ball of mass 250 g thrown at speed $8.0\,\text{m s}^{-1}$
   ii an electron moving at 1% of the speed of light

b Estimate the (magnitude of the) momentum of:
   i a charging elephant
   ii a scurrying mouse

Answer

a i $p = mv = 0.250\,\text{kg} \times 8.0\,\text{m s}^{-1} = 2.0\,\text{kg m s}^{-1}$
    ii $p = mv = 9.11 \times 10^{-31}\,\text{kg} \times 3.0 \times 10^{6}\,\text{m s}^{-1} = 2.7 \times 10^{-24}\,\text{kg m s}^{-1}$
b i $5000\,\text{kg} \times 8\,\text{m s}^{-1} = 4 \times 10^{4}\,\text{kg m s}^{-1}$
    ii $0.02\,\text{kg} \times 1\,\text{m s}^{-1} = 0.02\,\text{kg m s}^{-1}$

> **Exam tip**
> The electron mass and the speed of light are given in the data sheet, which is included in the specification (as Appendix 9) and printed at the end of each examination paper.

> **Exam tip**
> For estimates, any reasonable values are acceptable. Here it is assumed that an elephant has a mass of several tonnes and can charge with a speed similar to that of a sprinting athlete, whereas a mouse has a mass of less than 100 g and can run at a speed of around 1 metre per second.

The above examples illustrate the concept of momentum. The charging elephant has a large momentum, so stopping it or changing the direction of its motion would be difficult. On the other hand, it would require little effort to alter the state of the mouse's motion.

It is important to remember that momentum is a vector quantity and, as such, requires a direction in addition to the magnitude. Often, as in the examples above, the direction is implied; however, strictly speaking the direction should always be specified — for instance, by stating that the momentum of the elephant is $4 \times 10^{4}\,\text{kg m s}^{-1}$ due south. Information about the direction is very significant when studying problems in which moving bodies collide and changes in momentum take place.

Worked example

A rubber ball of mass 0.15 kg is dropped and attains a speed of $5.0\,\text{m s}^{-1}$ the instant it strikes the floor. It rebounds upwards with an initial speed of $4.0\,\text{m s}^{-1}$. Calculate:
a the momentum of the ball immediately before and after the rebound
b the change in momentum of the ball during the time it is in contact with the floor

Answer

a Just before hitting the ground:

$p_1 = 0.15\,\text{kg} \times 5.0\,\text{m s}^{-1} = 0.75\,\text{kg m s}^{-1}$ downwards

After the rebound:

$p_2 = 0.15\,\text{kg} \times 4.0\,\text{m s}^{-1} = 0.60\,\text{kg m s}^{-1}$ upwards

b If we assign a positive value to the upward velocity, then the downward velocity takes a negative sign.
So change in momentum of the ball $= p_2 - p_1$
$= (0.60\,\text{kg m s}^{-1}) - (-0.75\,\text{kg m s}^{-1})$
$= 1.35\,\text{kg m s}^{-1}$

> **Knowledge check 12**
> Calculate the momentum of a bullet of mass 50 g moving with a velocity of $220\,\text{m s}^{-1}$.

Mechanics

Principle of conservation of linear momentum

The **conservation of linear momentum** is a fundamental law of physics.

It is important that you fully understand the meaning of the expression 'system of interacting bodies' in the statement. It is evident that if two snooker balls collide and one of the balls is slowed down while the other speeds up, the momentum of each of the balls will change. However, the *vector sum* of the momentums of the balls after the collision will be the same as that immediately before the interaction.

The 'system' for the bouncing ball in the previous example is more difficult to imagine. The Earth is the second body in the system, but it is so massive compared with the ball that the change in its momentum due to the interaction will be imperceptible; nonetheless, the principle of conservation of momentum still applies.

The **conservation of linear momentum** states that in any system of interacting bodies the total momentum is conserved, provided that no resultant external force acts on the system.

Worked example

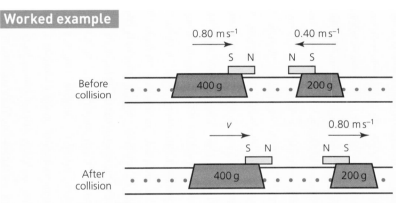

Figure 25

Two gliders moving towards each other on an air track 'collide' and then move apart, as shown in Figure 25. Initially the 400 g glider travels from left to right with a velocity of $0.80 \, \text{m s}^{-1}$ and the 200 g glider moves in the opposite direction at $0.40 \, \text{m s}^{-1}$. After the collision the smaller glider changes direction and moves back with a velocity of $0.80 \, \text{m s}^{-1}$.

a Calculate the velocity of the larger glider after the collision.

b If one of the magnets was reversed so that the gliders stuck together on impact, calculate the velocity of the combination.

Exam tip

It is useful to sketch a simple diagram showing the masses and velocities of the bodies before and after the interaction.

Remember that momentum is a vector. If you assign positive values for left to right motion, the velocities and momentum in the right for left direction must be negative.

Answer

a Let the rightward direction correspond to positive values, and let v denote the velocity of the larger glider after the collision. Then, for this system, initial momentum $= 0.400 \, \text{kg} \times 0.80 \, \text{m s}^{-1} + 0.200 \, \text{kg} \times (-0.40 \, \text{m s}^{-1}) = 0.240 \, \text{kg m s}^{-1}$
final momentum $= 0.400 \, \text{kg} \times v + 0.200 \, \text{kg} \times 0.80 \, \text{m s}^{-1}$
Conservation of momentum gives:
$0.240 \, \text{kg m s}^{-1} = 0.400 \, \text{kg} \times v + 0.200 \, \text{kg} \times 0.80 \, \text{m s}^{-1}$
so $v = +0.20 \, \text{m s}^{-1}$
Thus the larger glider moves with a velocity of $0.20 \, \text{m s}^{-1}$ to the right.

b $0.240 \, \text{kg m s}^{-1} = (0.400 \, \text{kg} + 0.200 \, \text{kg}) \times v_{\text{comb}}$
$v_{\text{comb}} = +0.40 \, \text{m s}^{-1}$
The combination of two gliders moves with a velocity of $0.40 \, \text{m s}^{-1}$ to the right.

Content Guidance

For oblique collisions in which the masses approach or separate along different lines, the principle of conservation of momentum can be applied to the *components* of the momentum in any direction. In the AS examination only one-dimensional collisions are considered. Other applications of the principle are covered in the further mechanics section in the student guide covering Topics 6–8 in this series.

Momentum and Newton's laws

Earlier in this section, you learned a version of **Newton's second law of motion** that applies only to the special case where a resultant force acts upon a body of fixed mass and causes the body to accelerate. The law can be stated more generally in terms of the change in momentum that occurs when a resultant force is applied to a body.

In symbols:

$$\Sigma F = \frac{\Delta p}{\Delta t}$$

(if consistent SI units are used).

This statement of Newton's second law encompasses situations in which the mass is indeterminate (e.g. interactions involving neutrinos) or continually changing (e.g. a stream of water striking a surface). However, in the examinations, Newton's second law will be applied only in situations where the mass is constant.

For a fixed mass:

$$\Delta F = \frac{\Delta(mv)}{\Delta t} = \frac{m\Delta v}{\Delta t} = ma$$

The above calculation confirms the validity of the equation used to represent Newton's second law when the force is applied to a fixed mass.

> **Newton's second law of motion** states that the rate of change in momentum is directly proportional to the resultant applied force.

> **Exam tip**
> Sometimes there are several forces acting on an object. It is essential that the *resultant* force on the body is determined before using the law.

Worked example

In an experiment to investigate Newton's second law, a glider is pulled along an air track by a falling mass, as shown in Figure 26.

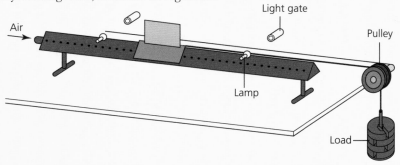

Figure 26

The time taken for the interrupter card to pass through each light gate is recorded, together with the time taken for the glider to travel between the gates. The results of one set of measurements are:
- mass of glider   400 g
- length of interrupter card   20.0 cm

26 Edexcel Physics

Mechanics

- time to cross first gate   0.316 s
- time to cross second gate   0.224 s
- time between gates   0.524 s

Use these results to calculate:

a the initial momentum of the glider
b the final momentum of the glider
c the resultant force acting on the glider

The mass on the load is 22.0 g.

d Comment on how this compares with the resultant force on the glider calculated in part (c).

Answer

a initial velocity = $\dfrac{0.200\,\text{m}}{0.316\,\text{s}}$ = 0.633 m s^{-1}

initial momentum = 0.400 kg × 0.633 m s^{-1} = 0.253 kg m s^{-1}

b final velocity = $\dfrac{0.200\,\text{m}}{0.224\,\text{s}}$ = 0.893 m s^{-1}

final momentum = 0.400 kg × 0.893 m s^{-1} = 0.357 kg m s^{-1}

c resultant force = $\dfrac{\text{change in momentum}}{\text{time}}$

$= \dfrac{(0.357 - 0.253)\,\text{kg m s}^{-1}}{0.524\,\text{s}} = 0.198\,\text{N}$

d The force pulling the glider is the weight of the load, i.e.

0.0220 kg × 9.81 m s^{-2} = 0.216 N

This suggests that there must be a resistive force of 0.018 N acting on the glider, so that the resultant force is 0.216 N − 0.018 N = 0.198 N as found in part (c).

Newton's third law of motion is also closely related to the principle of conservation of momentum. Consider two boys on skateboards facing each other on a horizontal surface (Figure 27).

Figure 27

Exam tip

It is important to realise that this law does not contradict the law of conservation of momentum. Although the application of a resultant force may change the momentum of individual masses, the total momentum change for a system of two or more masses will always be zero.

Content Guidance

If the boys push each other, they will both move backwards away from each other. Using the principle of conservation of momentum, the change in the momentum of each boy must be equal and opposite. Because the time, Δt, taken for both pushes must be the same, it follows that:

the force of boy A on boy B = the rate of change in momentum of boy B

$= \dfrac{\Delta p_B}{\Delta t}$

the force of boy B on boy A = the rate of change in momentum of boy A

$= \dfrac{\Delta p_A}{\Delta t}$

As $\Delta p_B = -\Delta p_A$ it follows that the force exerted by boy A on boy B is equal and opposite to that exerted by boy B on boy A.

This shows that Newton's third law is an example of the principle of conservation of momentum.

Turning forces

When a force is applied to an object that is not a point mass it may rotate. The turning effect of a force is called its **moment**. The moment of a force depends on the size of the force and the perpendicular distance of its line of action from the axis of rotation.

The units of a moment are newton metres (N m). This is not the same as the unit of energy (the joule) because the force causing the moment does not actually move. The moment has a direction of rotation, clockwise or anticlockwise, and can be treated in a similar manner to a vector.

Principle of moments

When a body is acted upon by a number of forces, the resultant moment can be calculated by adding the total clockwise and anticlockwise moments.

If the body is in equilibrium there can be no resultant moment acting upon it; that means that the sum of the clockwise moments must equal the sum of the anticlockwise moments. This is known as the **principle of moments**.

> The **moment** of a force is the product of the force and the perpendicular distance of the line of action of the force from the point of rotation.

> **Knowledge check 13**
> Calculate the moment of a 12 N force acting at a perpendicular distance of 25 cm from the axis of rotation.

> The **principle of moments** states that if a system is in equilibrium the sum of the clockwise moments about any point must equal the sum of the anticlockwise moments about that point.

> **Exam tip**
> Always indicate the direction of rotation by assigning a positive sign to the clockwise moments and a negative sign to the anticlockwise moments.

Worked example

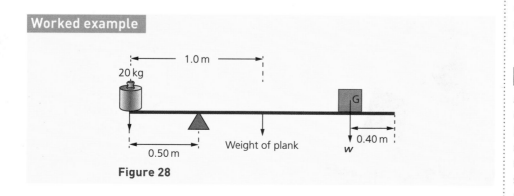

Figure 28

In Figure 28, a box is balanced on a uniform plank of length 2.0 m with a pivot 50 cm from one end, using a 20 kg mass placed at the end closest to the pivot. If the centre of gravity of the box is 40 cm from the other end of the plank and the mass of the plank is 5.0 kg, calculate:

a the weight, W, of the box

b the upward push, N, of the pivot on the plank

Answer

a Taking moments about the pivot:
anticlockwise moments = 20 kg × 9.8 m s^{-2} × 0.50 m = 98 N m
clockwise moments = W × 1.1 m + 5.0 kg × 9.8 m s^{-2} × 0.50 m
For equilibrium:
sum of the clockwise moments = sum of the anticlockwise moments
W × 1.1 m + 5.0 kg × 9.8 m s^{-2} × 0.50 m = 98 N m
W = 67 N

b For equilibrium the resultant force acting on the system is zero:
N − weight of (20 kg mass + box) = 0
N = 20 kg × 9.8 N kg^{-1} + 5 kg × 9.8 N kg^{-1} + 67 N
   = 196 N + 49 N + 67 N
   = 312 N

> **Exam tip**
>
> Always start your answer by stating 'sum of the clockwise moments = sum of the anticlockwise moments'.

Summary

After studying this topic, you should be able to:
- state Newton's first law of motion
- draw free-body force diagrams for a body in a state of equilibrium
- state Newton's second law of motion for bodies of fixed mass
- determine the resultant force acting on a body, and use $F = ma$ to calculate its acceleration
- understand the concept of gravitational field strength, and the difference between mass and weight
- state Newton's third law of motion
- describe the nature and direction of pairs of forces on two interacting bodies
- define momentum and use the principle of conservation of linear momentum
- understand the meaning of the moment of a force and apply the principle of moments to systems in equilibrium

---

Work, energy and power

Work and energy are interrelated. When work is done on a body, the body will gain energy, and a body can transfer energy to do work. Power is the rate of doing work.

Work

- Work is done when the point of application of a force is moved.
- $\Delta W = F\Delta x$ is the equation defining the work done when the point of application of a force, F, is moved a distance Δx.
- 1 joule (J) of work is done when the point of application of a force of 1 newton (N) is moved through a distance of 1 metre (m).

> **Work** done equals the product of the force *times* the distance moved *in the direction of the applied force*.

Content Guidance

> **Worked example**
>
> Calculate the work done in lifting a load of 8.0 kg onto a shelf of height 1.6 m.
>
> **Answer**
>
> force needed to raise the load, $F = mg = 8.0 \text{ kg} \times 9.8 \text{ m s}^{-2} = 78 \text{ N}$
>
> work done = $F\Delta x = 78 \text{ N} \times 1.6 \text{ m} = 125 \text{ J}$

> **Exam tip**
>
> Most questions involving objects to be lifted give the mass of the object. It is a common error to use this as the weight. Remember always to multiply the mass by 9.8 m s^{-2} in order to find the weight of a body.

The definition of work requires that the distance moved is in the direction of the point of application of the force. In some cases a body will move in a different direction from the applied force. Consider a sleigh being pulled across the snow by a piece of rope (Figure 29).

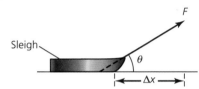

Figure 29

The direction of movement is at an angle θ to the applied force, so the work done will be given by:

$$\Delta W = F \cos \theta \times \Delta x = F\Delta x \cos \theta$$

You might think from the above expression that work is a vector quantity. However, if the sleigh is being pulled along a curved path, work is continually being done on it in ever-changing directions, and so work is a scalar quantity.

Energy

- **Energy** is a scalar quantity, measured in joules.
- Energy cannot be created or destroyed, but it can be converted to another form or transferred as work.

> **Energy** is the capacity to do work.

The relationship between work and energy and the law of conservation of energy are fundamental principles in physics at all levels. You will probably have considered the energy changes in simple systems such as a light bulb connected to a battery (chemical energy → electrical energy → thermal and light energy) or even the more complex system of a gas-fired power station (chemical energy in the gas → thermal energy → kinetic energy of the turbine → electrical energy in the generator). In this part of the course you will be required to look more deeply into the mechanical forms of energy: **potential energy** and **kinetic energy**.

Gravitational potential energy (GPE)

If a mass m is raised from the floor through a height Δh, the work done on the mass is:

$$\Delta W = mg\Delta h$$

The mass is now in a position to do work by falling back to the floor. The mass has potential energy due to its position, known as **gravitational potential energy (GPE)**.

> **Gravitational potential energy** is the ability of a body to do work by virtue of its position in a gravitational field.

30 Edexcel Physics

Mechanics

The change in gravitational potential energy when the mass is raised is:

$\Delta \text{GPE} = mg\Delta h$

Elastic potential energy (EPE)

A stretched rubber band can do work when it is released. The band is therefore storing energy while it is in its deformed state. This energy is called **elastic potential energy (EPE)** or **elastic strain energy**. Any elastically deformed object will store EPE. Clockwork springs, loaded crossbows and catapults (slingshots) are all examples of systems storing elastic potential energy.

Elastic strain energy is covered in more detail in the section on solid materials in the second student guide covering Topics 4 and 5 in this series.

Kinetic energy (KE)

If a cyclist freewheeling along a level road applies the brakes, the frictional resistive force between the brake blocks and the wheel will bring the bicycle to rest over a distance depending on its speed, the nature of the brakes etc. The moving bicycle and rider will therefore be doing work against the resistive forces. The moving bicycle must have some energy to do this work; this energy is called **kinetic energy** (KE).

The kinetic energy of a body of mass, m, moving with a velocity, v, is given by the expression:

$\text{KE} = \tfrac{1}{2}mv^2$

In the case of the cyclist, the kinetic energy is transferred to the brake blocks as thermal energy (and sound) that is dissipated into the environment.

For a body in free fall, the conservation of energy means that any loss in GPE equals the gain in KE (provided air resistance is ignored).

> **Kinetic energy** is the ability of a body to do work by virtue of its motion.

> **Knowledge check 14**
>
> Describe the energy changes during a single cycle of an oscillating pendulum.

Worked example

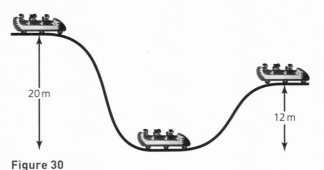

Figure 30

The car on a roller-coaster ride falls through a height of 20 m from rest at the top of the first peak to the bottom of the first dip. At the top of the second peak, 12 m above the dip, it is travelling at $12\,\text{m}\,\text{s}^{-1}$. The mass of the car and passengers is 2000 kg (Figure 30).

a Show that the speed of the car in the dip is about $20\,\text{m}\,\text{s}^{-1}$.

b Show that the work done by the car against the resistive forces between the two peaks is about 10 kJ.

> **Exam tip**
>
> It is a common error for students to apply the equation of motion $v^2 = u^2 + 2as$ to determine v. Although this may give the correct value of v, it is an error of physics (the '*suvat*' equations only apply to uniform motion in a straight line) and will be awarded zero marks in the AS examination.

Content Guidance

Answer

a loss in GPE = gain in KE

$mg\Delta h = \frac{1}{2}mv^2$

$v = \sqrt{2gh} = \sqrt{2 \times 9.8\,\text{m s}^{-2} \times 20\,\text{m}} = 19.8\,\text{m s}^{-1}$

b loss in GPE = gain in KE + work done against resistive forces
$2000\,\text{kg} \times 9.8\,\text{m s}^{-2} \times (20\,\text{m} - 12\,\text{m}) = \frac{1}{2} \times 2000\,\text{kg} \times (12\,\text{m s}^{-1})^2 + \Delta W$
$\Delta W = 157\,\text{kJ} - 144\,\text{kJ} = 13\,\text{kJ} \approx 10\,\text{kJ}$

> **Exam tip**
> In a 'show that' question the answer should be given to at least one more significant figure than the approximate value shown in the question.

Power

The transfer of **power** depends not only on the energy converted, but also on how quickly it is transferred. Power can be represented by the expressions:

$$P = \frac{\Delta W}{\Delta t} = \frac{\Delta E}{\Delta t}$$

Power is measured in watts, W (J s^{-1}).

> **Power** is the rate of doing work, or transferring energy.

Worked example

A man of mass 75 kg runs up a flight of stairs with 20 steps each of height 20 cm in a time of 7.0 s. Calculate the average power expended by the man.

Answer

work done = $mg\Delta h$ = 75 kg × 9.8 m s^{-2} × (20 × 0.20 m) = 2940 J

power = $\dfrac{2940\,\text{J}}{7.0\,\text{s}}$ = 420 W

The power of an object moving at constant velocity against a uniform resistive force can be found from the product of the force and the velocity:

$$P = \frac{\Delta W}{\Delta t} = \frac{F\Delta x}{\Delta t} = F \times \frac{\Delta x}{\Delta t} = Fv$$

> **Exam tip**
> In most cases, e.g. riding a bicycle, resistive forces may vary, so average values are generally stated.

Efficiency

Earlier in this section the energy conversions for a light bulb and a power station were given. In both these cases much of the initial energy input is not converted to the desired output. For a filament lamp most of the electrical energy is converted to thermal energy, leaving a relatively small percentage as light. Similarly, throughout the conversions in the power station, and particularly in the turbines, a large amount of thermal energy is 'lost' to the surroundings. The **efficiency** of a system relates to how much useful output is gained compared with the energy put into the system.

$$\text{efficiency} = \frac{\text{useful energy (power) output}}{\text{energy (power) input}} \times 100\%$$

Worked example

A mobility scooter is powered by a 300 W electric motor. It moves with a speed of $2.5\,\mathrm{m\,s^{-1}}$ against an average resistive force of 100 N. Calculate the output power, and hence the efficiency of the buggy.

Answer

$$\text{output power} = F \times v = 100\,\mathrm{N} \times 2.5\,\mathrm{m\,s^{-1}} = 250\,\mathrm{W}$$

$$\text{efficiency} = \frac{250\,\mathrm{W}}{300\,\mathrm{W}} \times 100\% = 83\%$$

Knowledge check 15

Why is it not possible for machines to work with greater than 100% efficiency?

Summary

After studying this topic, you should be able to:
- calculate the work done when the point of application of a force is moved by a given distance
- describe energy transformations for a variety of applications
- calculate the kinetic and gravitational potential energies of bodies, and use these to determine changes in velocity or position of the body
- determine the power and efficiency of energy transformations

Content Guidance

Electric circuits

An **electric current** is a flow of charge, i.e. a movement of charge carriers such as electrons and ions. Direct current (DC) flows in one direction only. In order to make charges move, work must be done. A 'voltage', or more correctly a **potential difference**, between two points is needed to provide this energy. When a current flows in a circuit, it will encounter a **resistance** to the flow.

The quantities are represented by the following equations: note that Q = charge, in coulombs (C); W = energy or work done transferring a charge, in joules (J)

- $I = \dfrac{\Delta Q}{\Delta t}$ unit: ampere (A)
- $V = \dfrac{W}{Q}$ unit: volt (V)
- $R = \dfrac{V}{I}$ unit: ohm (Ω)

> **Current**, I = rate of flow of charge
>
> **Potential difference**, V = work done in transferring 1 coulomb of charge
>
> **Resistance**, R = potential difference divided by current

Worked example

a How much charge flows through a filament lamp in 1 minute when the current is 500 mA?

b Given that the electron charge is 1.6×10^{-19} C, how many electrons flow during this time?

c What is the potential difference across the filament if 60 J of energy is emitted by the filament?

d Calculate the resistance of the filament.

Answer

a $\Delta Q = I\Delta t = (500 \times 10^{-3}\,\text{A}) \times 60\,\text{s} = 30\,\text{C}$

b number of electrons = $\dfrac{\text{total charge}}{\text{charge on each electron}}$

$= \dfrac{30\,\text{C}}{1.6 \times 10^{-19}\,\text{C}} = 1.9 \times 10^{20}$ electrons

c $V = \dfrac{W}{Q} = \dfrac{60\,\text{J}}{30\,\text{C}} = 2.0\,\text{V}$

d $R = \dfrac{V}{I} = \dfrac{2.0\,\text{V}}{0.5\,\text{A}} = 4.0\,\Omega$

> **Exam tip**
>
> The expressions $V = IR$ and $I = V/R$ are useful arrangements of the definition of resistance, but they are not accepted as definitions of potential difference and current.

Series circuits

A series circuit is a circuit in which the components are connected in sequence, one after the other (Figure 31).

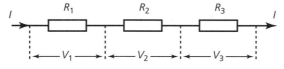

Figure 31

34 Edexcel Physics

Electric circuits

In all circuits, the laws of conservation of charge and conservation of energy must apply. In Figure 31, three resistors are connected in series. For charge to be conserved, the rate of flow of charge must be the same through each resistor. It follows that *in series circuits the current I must be the same in all components*.

For conservation of energy to hold, the total energy transferred to the resistors must equal the sum of the energies transferred in each component. Therefore, the total work done per coulomb transferred will equal the sum of the work done per coulomb transferred in each of the resistors. Since work done per coulomb transferred is the same as potential difference, we have:

$V_{total} = V_1 + V_2 + V_3$

Using $V = IR$ from the definition of resistance and the fact that I is the same throughout the circuit, the above equation becomes $IR_{total} = IR_1 + IR_2 + IR_3$. Then, cancelling out I gives the total resistance in the series circuit as:

$R_{total} = R_1 + R_2 + R_3$

Because the current is the same at all parts of a series circuit, an ammeter connected in series will give the same reading no matter where it is placed relative to the resistors. For the ammeter itself to have little effect on the overall resistance of the circuit, it needs to have an extremely small resistance.

Parallel circuits

In a parallel circuit, the ends of each resistor are connected to points that have the same potential difference (Figure 32).

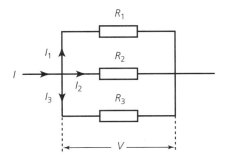

Figure 32

From conservation of charge it follows that the total charge entering a junction must be the same as that leaving the junction, so:

$Q = Q_1 + Q_2 + Q_3$

where Q denotes the total charge flowing through the circuit and Q_1, Q_2, Q_3 are the amounts of charge flowing through the separate resistors.

The rates of flow of charge into and away from the junction must then also be equal, giving:

$I = I_1 + I_2 + I_3$

Content Guidance

The law of conservation of energy requires that the ratio W/Q be the same for all three resistors; in other words, the potential difference must be the same across all three resistors: $V_1 = V_2 = V_3 = V$. Using the definition of resistance in the rearranged form $I = V/R$, we get:

$$\frac{V}{R} = \frac{V}{R_1} + \frac{V}{R_2} + \frac{V}{R_3}$$

and hence:

$$\frac{1}{R} = \frac{1}{R_1} + \frac{1}{R_2} + \frac{1}{R_3}$$

Worked example

1 Calculate the total resistance of each of the three circuits shown in Figure 33.

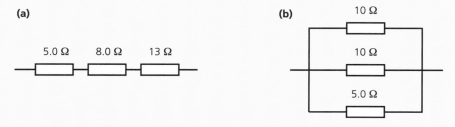

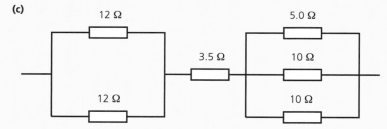

Figure 33

2 A potential difference of 6.0 V is applied across the ends of circuit (c) above. Calculate:
   a the total current in the circuit
   b the potential difference across the 5.0 Ω resistor
   c the current in the 5.0 Ω resistor

Answer

1 a Three resistors in series, so $R = 5.0\,\Omega + 8.0\,\Omega + 13\,\Omega = 26\,\Omega$
  b Three resistors in parallel, so:

$$\frac{1}{R} = \frac{1}{10\,\Omega} + \frac{1}{10\,\Omega} + \frac{1}{5.0\,\Omega} = 0.40\,\Omega^{-1} \text{ and hence } R = 2.5\,\Omega$$

  c We need to find the resistance of each of the two parallel networks and then add the three resistances in series.

> **Exam tip**
>
> For parallel circuits, the following points are worth remembering:
> - The total resistance is always less than that of the smallest resistor.
> - If all the resistors are equal, the total resistance will equal the resistance of each resistor divided by the number of resistors (e.g. two 10 Ω resistors in parallel give a total resistance of 5 Ω; i.e. $1/R = 1/10 + 1/10 \Rightarrow 1/R = 1/5 \Rightarrow R = 5\,\Omega$).
> - At the end of a parallel resistance calculation, don't forget to take the reciprocal of the $1/R$ value to find R.

Electric circuits

For the parallel circuit on the left:

$\frac{1}{R} = \frac{1}{12\,\Omega} + \frac{1}{12\,\Omega}$, which gives $R = 6.0\,\Omega$

For the parallel circuit on the right:

$\frac{1}{R} = \frac{1}{5.0\,\Omega} + \frac{1}{10\,\Omega} + \frac{1}{10\,\Omega}$, which gives $R = 2.5\,\Omega$ (see answer b)

The total resistance is $6.0\,\Omega + 3.5\,\Omega + 2.5\,\Omega = 12\,\Omega$

2 a $I = \frac{V}{R} = \frac{6.0\,\text{V}}{12\,\Omega} = 0.50\,\text{A}$

 b The potential difference across the $5.0\,\Omega$ resistor is the same as the potential difference across the whole right-hand side circuit of three resistors in parallel. Now, the current flowing through the right-hand side parallel circuit is the same as the total current that flows in the entire circuit; so, using $V = IR$ for the right-hand side parallel circuit, we get:

$\frac{\text{potential}}{\text{difference}} = \text{total current} \times \frac{\text{resistance of right-hand}}{\text{side parallel circuit}}$

$V = 0.50\,\text{A} \times 2.5\,\Omega = 1.25\,\text{V}$

 c $I = \frac{V}{R} = \frac{1.25\,\text{V}}{5.0\,\Omega} = 0.25\,\text{A}$

Voltmeters measure the potential difference across components such as resistors. Therefore they must be connected in parallel with the components. Some current will be diverted from the circuit by the voltmeter. For this amount of current to be negligible, the voltmeter needs to have a very large resistance.

Electrical energy and power

You have seen that in order for a current to flow, work needs to be done on the charge carriers. This is expressed in the definition of potential difference, which says that $V = W/Q$ or, more accurately, since the denominator is 'amount of charge transferred', $V = W/\Delta Q$. Rearranging and combining with the formula $I = \Delta Q/\Delta t$ gives:

$W = V\Delta Q = VI\Delta t$

When a current passes through a resistor, the work done converts electrical energy to thermal energy in the resistor. For a steady current, the **electrical energy** transferred to a circuit in time t is therefore:

$E = VIt$

Worked example

The heating element of the rear window of a car has a resistance of $2.0\,\Omega$ and is connected to a $12\,\text{V}$ battery. How much energy does it give off in 20 minutes?

Answer

$I = \frac{V}{R} = \frac{12\,\text{V}}{2.0\,\Omega} = 6.0\,\text{A}$

$W = VIt = 12\,\text{V} \times 6.0\,\text{A} \times (20 \times 60\,\text{s}) = 86\,\text{kJ}$

Knowledge check 16

Calculate the total resistance of three $15\,\Omega$ resistors connected in parallel, in series with a pair of $18\,\Omega$ resistors that are also connected in parallel.

Electrical energy is the work done by a potential difference, V, when a charge ΔQ is transferred. The energy transferred is the product $V\Delta Q$.

Content Guidance

In the mechanics section, **power** was defined as the rate of doing work:
$$P = \frac{\Delta W}{\Delta t}$$
Hence, for a steady current:
$$P = \frac{VIt}{t} = VI$$
Using the definition of resistance, $R = V/I$, this can also be written as:
$$P = I^2R = \frac{V^2}{R}$$

> **Electrical power** is the rate at which electrical energy is dissipated in a circuit. It is usually defined by the equation $P = VI$.

Worked example

The power from a 10 MW wind farm is transmitted at a potential difference of 330 000 V.

a Calculate the current produced when the farm is generating at its maximum capacity.

b The electricity is transferred to a sub-station through overhead cables of resistance 100 Ω. Calculate the power loss in the cables.

c A student suggested that it would be safer and more efficient to transmit the power at 100 000 V. Determine the power that would be lost in the cables at this voltage of transmission, and comment on the student's suggestion.

Answer

a $P = VI \Rightarrow 10 \times 10^6\,\text{W} = 330\,000\,\text{V} \times I \Rightarrow I = \dfrac{10 \times 10^6\,\text{W}}{330 \times 10^3\,\text{V}} = 30\,\text{A}$ (to 2 s.f.)

b $\Delta P = I^2 R = (30\,\text{A})^2 \times 100\,\Omega = 90\,\text{kW}$

c At a transmission voltage of 10^5 V, the current I would be:
$$\frac{10^7\,\text{W}}{10^5\,\text{V}} = 100\,\text{A}$$
So the power loss would be:
$$\Delta P = I^2 R = (100\,\text{A})^2 \times 100\,\Omega = 1.0\,\text{MW}$$

At a lower voltage the current will be larger and, since the power loss in the cables is dependent on the square of the current, much more energy will be transferred to the surroundings when the transmission voltage is lower. For this reason, the national grid transmits at as high a voltage as possible.

The dangers of high voltages are a problem, of course, and the voltage is stepped down for domestic use. However, even from a safety perspective there is little to be gained by dropping the transmission voltage from 330 000 V to 100 000 V.

> **Knowledge check 17**
>
> Calculate the energy given out by an electric kettle element of resistance 20 Ω in 5.0 minutes. The mains voltage is 230 V.

> **Ohm's law** states that the current in a metallic conductor at constant temperature is directly proportional to the potential difference applied to its ends.

> **Exam tip**
>
> A common error is to take one of the equations for the definition of resistance ($R = V/I$, $I = V/R$ or $V = IR$) as the statement of Ohm's law. This is not valid, because these equations hold in all conditions, whereas Ohm's law applies only to metals at constant temperature.

Ohm's law

Ohm's law is a statement on how the current in a metallic conductor relates to the potential difference across the ends of the conductor.

Some non-metallic components can behave in the same way as metals at a fixed temperature. Such conductors, in which the current is proportional to the voltage, are said to be **ohmic**.

Electric circuits

The characteristics of a number of different conductors can be studied by applying a range of potential differences across them, measuring the corresponding currents and plotting the associated I–V graphs. This may be achieved by using ammeters and voltmeters or, if the precision of the readings is not important, the data may be fed into a computer interface and the graphs displayed electronically (Figure 34).

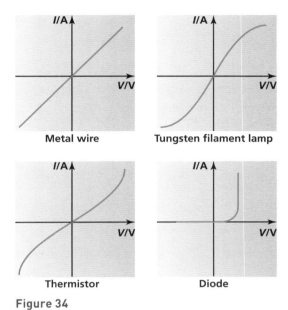

Figure 34

For ohmic conductors, I is always proportional to V so that the ratio of potential difference to current (V/I) is constant. This means that the resistance of an ohmic conductor is the same at all values of potential difference and current. In other, non-ohmic, components the resistance is variable. From the graphs in Figure 34, it can be seen that in the filament lamp, the ratio V/I (i.e. the resistance) increases as the current grows. For the thermistor, on the other hand, resistance decreases with current. The resistance in these components is affected by the temperature, a phenomenon that will be discussed later. Diodes are designed to allow the current to flow in one direction only — they have an extremely high resistance when 'reverse biased', and become conducting only after a small forward potential difference is applied.

Resistivity

If asked whether copper wire has lower resistance than nichrome wire, it is tempting to answer yes, as copper is used for connecting leads and nichrome for heating elements. However, if you measured the resistance of a 20 m length of copper wire of diameter 0.5 mm, the value would turn out to be greater than the resistance of a 10 cm-long nichrome rod of diameter 1 cm.

Simple measurements show that the resistance of a piece of wire is directly proportional to its length and inversely proportional to its cross-sectional area:

$$R = \frac{\rho l}{A}$$

Exam tip

The resistance of a conductor depends not only on the material from which it is made, but also on its dimensions. Resistivity is a property of a material, and will be constant provided the temperature of the material does not change.

Content Guidance

The constant of proportionality, ρ, is dependent on the material from which the wire is made and is known as the **resistivity** of the material.

$$\text{resistivity} = \frac{\text{resistance} \times \text{cross-sectional area}}{\text{length}}$$

$$\rho = \frac{RA}{l} \quad \text{unit: } \Omega\,\text{m}$$

- resistivity is a property of the material
- resistivity varies with temperature

Worked example

a Calculate the length of nichrome wire of diameter 1.2 mm that is needed to make a 5.0 Ω resistor. The resistivity of nichrome is $1.1 \times 10^{-6}\,\Omega\,\text{m}$.

b

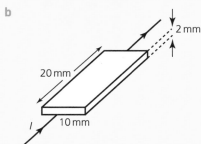

Figure 35

A graphite chip has dimensions 20 mm × 10 mm × 2 mm and resistivity $7 \times 10^{-5}\,\Omega\,\text{m}$. Calculate the resistance of the chip for a current in the direction shown (Figure 35).

Answer

a $l = \dfrac{RA}{\rho} = \dfrac{5.0\,\Omega \times \pi(0.6 \times 10^{-3}\,\text{m})^2}{1.1 \times 10^{-6}\,\Omega\,\text{m}} = 5.1\,\text{m}$

b $R = \dfrac{\rho l}{A} = \dfrac{(7 \times 10^{-5}\,\Omega\,\text{m}) \times (20 \times 10^{-3}\,\text{m})}{(10 \times 10^{-3}\,\text{m}) \times (2 \times 10^{-3}\,\text{m})} = 0.07\,\Omega$

Core practical 2

Determining the electrical resistivity of materials

Core practical 2 requires you to determine the resistivity of a metal by measuring the resistance of a range of lengths of wire and the diameter of the wire. Full details of how these measurements are taken, precautions taken and how the resistivity is found using a graphical method may be required for the examinations.

Knowledge check 18

Determine the resistivity of copper, if a 2.0 m length of copper wire of diameter 0.50 mm has a resistance of 1.8 Ω.

Electric circuits

Potential dividers

In DC circuits it is often necessary to apply a potential difference across a component that is less than the voltage of the supply. It is quite simple to use a pair of resistors connected in series to divide the voltage by 'sharing' it between the resistors (Figure 36).

The ratio of the voltages across the two resistors can be written as:

$$\frac{V_1}{V_2} = \frac{IR_1}{IR_2} = \frac{R_1}{R_2}$$

since I is the same in each resistor.

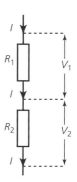

Figure 36

It is common to show a potential divider giving a required output voltage V_{out} from a supply of voltage V_{in}. In this case, $V_{out} = V_2$ and $V_{in} = V_1 + V_2$, which leads to:

$$\frac{V_{out}}{V_{in}} = \frac{R_2}{R_1 + R_2}$$

and hence:

$$V_{out} = \frac{R_2}{R_1 + R_2} \times V_{in}$$

> **Exam tip**
>
> The principle of a potential divider is that the ratio of the voltages across the resistors connected in series is equal to the ratio of their resistances.

Worked example

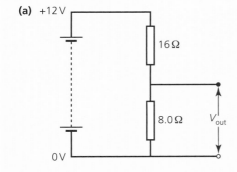
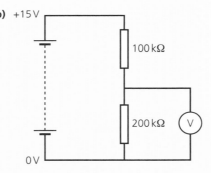

Figure 37

a Calculate the output voltage across the $8.0\,\Omega$ resistor in Figure 37(a).

b i Calculate the output voltage in the circuit in Figure 37(b).

  ii A voltmeter of resistance $200\,\text{k}\Omega$ is connected across the output. What reading will it give for the output voltage?

Answer

a $V_{out} = \dfrac{8.0\,\Omega}{16\,\Omega + 8.0\,\Omega} \times 12\,\text{V} = 4.0\,\text{V}$

b i $V_{out} = \dfrac{200\,\text{k}\Omega}{100\,\text{k}\Omega + 200\,\text{k}\Omega} \times 15\,\text{V} = 10\,\text{V}$

  ii The voltmeter and the $200\,\text{k}\Omega$ resistor in parallel will give a combined resistance of $100\,\text{k}\Omega$. Hence:

  $V_{out} = \dfrac{100\,\text{k}\Omega}{100\,\text{k}\Omega + 100\,\text{k}\Omega} \times 15\,\text{V} = 7.5\,\text{V}$

Content Guidance

Note: this is an example of how using an instrument to measure the value of a quantity can affect that measured value. In most cases, especially with digital meters, the voltmeter has such a high resistance that it takes virtually no current from the network, thus leaving the voltage almost unaffected.

Potentiometers

A potentiometer can be used to obtain a continuously variable output voltage (Figure 38).

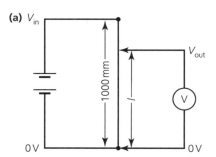
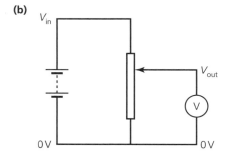

Figure 38

> **Exam tip**
>
> Potentiometers are basically potential dividers in which the resistors can be continuously varied.

> **Knowledge check 19**
>
> A 12 V battery is connected to the ends of a 1.00 m length of nichrome wire. Calculate the potential difference across 40 cm of the wire.

The simplest potentiometer consists of a uniform length of resistance wire stretched along a metre rule (Figure 38(a)). For uniform wire (where the cross-sectional area is constant), the resistance is proportional to the length. Thus, the output voltage for such a simple potentiometer of length l is given by:

$$V_{out} = \frac{l}{1.000\,\text{m}} \times V_{in}$$

Most practical potentiometers are more compact and use a sliding contact to vary the length of the output section along a wire coil or around a circular conductor (Figure 38(b)).

Applications of potentiometers include heat- or light-controlled switches. You have met thermistors, which have resistances that change with temperature. If a thermistor or a light-dependent resistor (LDR) is connected in series with a resistor, the output voltage (across the resistor) can be controlled by temperature or light variations (Figure 39).

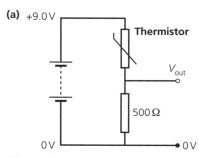
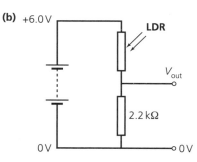

Figure 39

The resistor is chosen so that an alarm will be triggered at a specific output voltage. Alternatively, a voltmeter across the output can be calibrated according to temperature or luminous intensity.

Electric circuits

Worked example

In the circuit shown in Figure 39(a), the resistance of the thermistor is $2.0\,\text{k}\Omega$ at 20°C and $400\,\Omega$ at 50°C. Calculate the output voltage at each of these temperatures.

Answer

At 20°C:
$$V_{\text{out}} = \frac{500\,\Omega}{2000\,\Omega + 500\,\Omega} \times 9.0\,\text{V} = 1.8\,\text{V}$$

At 50°C:
$$V_{\text{out}} = \frac{500\,\Omega}{400\,\Omega + 500\,\Omega} \times 9.0\,\text{V} = 5.0\,\text{V}$$

Electromotive force

At the beginning of this section you were given a definition of potential difference in terms of the work done, or energy converted, within a circuit or component. The idea that a source of electrical energy is needed before a current can flow was also introduced.

The term **electromotive force**, commonly written as emf (or e.m.f.) is always used in relation to devices that convert energy from other forms into electrical energy.

$$\varepsilon = \frac{\Delta E}{\Delta Q} \qquad \text{unit: volt (V)}$$

Examples of electrical sources include:
- dynamo — converts kinetic energy into electrical
- battery — converts chemical energy into electrical
- solar cell — converts radiant (light) energy into electrical

By the law of conservation of energy, the chemical energy converted per coulomb of charge in the battery in Figure 40 must be equal to that transformed into thermal energy and light in the lamp and into kinetic energy in the motor.

emf of battery = potential difference across lamp + potential difference across motor

> The **emf** of a generator is the energy converted into electrical energy per coulomb of charge produced.

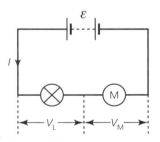

Figure 40

Internal resistance

Unfortunately, not all of the chemical energy in the battery shown above will be transferred to the lamp and the motor. Some of the energy is used to 'push' the charges through the cells of the battery. These cells have a resistance between the electrodes. This is called the **internal resistance** of the cell and is denoted by r (Figure 41).

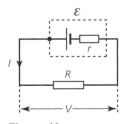

Figure 41

> **Exam tip**
>
> Electromotive force is only used for devices that generate electricity (e.g. batteries and dynamos). In all other cases the 'voltage' across a component is called the potential difference.

Content Guidance

Applying the law of conservation of energy as before, we have:

emf of cell = potential difference across resistor + work done per coulomb within the cell

$\varepsilon = V + \Delta V$

It follows that the **terminal potential difference** $V (= IR)$ will be less than the emf of the cell, and the difference ΔV, i.e. the 'lost volts', will equal Ir:

$\varepsilon = IR + Ir$

To calculate the terminal potential difference, it is useful to remember two rearrangements of the above equations:

- $V = \varepsilon - Ir$
- $I = \dfrac{\varepsilon}{R + r}$

> **Exam tip**
>
> The equation $V = \varepsilon - Ir$ is useful in the practical determination of the internal resistance of a cell. If the potential difference across the cell is measured for a range of currents, a graph of V against I results in a straight line of gradient $-r$ with the intercept on the y-axis equalling the emf of the cell.

Worked example

In daylight, a solar cell has an emf of 2.4 V and an internal resistance of 20 Ω. Calculate the terminal potential difference when the cell is connected to a resistor of resistance:

a 20 Ω
b 180 Ω
c 980 Ω

Answer

a $I = \dfrac{2.4\,\text{V}}{20\,\Omega + 20\,\Omega} = 0.060\,\text{A} \Rightarrow V = IR = 0.060\,\text{A} \times 20\,\Omega = 1.2\,\text{V}$

b $I = \dfrac{2.4\,\text{V}}{180\,\Omega + 20\,\Omega} = 1.2 \times 10^{-2}\,\text{A} \Rightarrow V = 1.2 \times 10^{-2}\,\text{A} \times 180\,\Omega = 2.16\,\text{V}$

c $I = \dfrac{2.4\,\text{V}}{980\,\Omega + 20\,\Omega} = 2.4 \times 10^{-3}\,\text{A} \Rightarrow V = 2.4 \times 10^{-3}\,\text{A} \times 980\,\Omega = 2.35\,\text{V}$

> **Exam tip**
>
> Check that using $V = \varepsilon - Ir$ gives the same result.

The above worked example shows that the terminal potential difference approaches the emf for external resistances that are much greater than the internal resistance. For extremely large external resistances, such as those of digital voltmeters, the current, and therefore the value of Ir, is virtually zero; thus the voltage across the terminals equals the emf of the cell.

For a battery of cells connected together in series, the total emf will equal the sum of the individual emfs, but the internal resistance of the battery will also be the sum of the component resistances.

For a number of identical cells connected in parallel, the emf will be the same as that of each cell, but the internal resistance will be reduced.

Electric circuits

Core practical 3

Determining the emf and internal resistance of an electrical cell

Core practical 3 requires you to determine the emf and internal resistance of an electrical cell by measuring the current and potential difference for a range of values of external resistance. Full details of how these measurements are taken, precautions taken and how the emf and internal resistance are found from a potential difference–current graph may be required for the examinations.

Factors affecting current in conductors

It was stated at the start of this section that an electric current is a flow of charge pushed around a circuit by a potential difference, and that work is done against the resistance of the conducting medium. You will need to have a more detailed knowledge of the mechanics of this process at an atomic level.

For *metallic* conductors, the charge is carried by **delocalised** or **free electrons**. These electrons are not fixed to specific atoms and are free to move around the atomic lattice. At room temperature, they have thermal energy and move at high speeds in a random manner, similar to gas particles in the atmosphere. Typically, there is about one free electron for every atom. When a potential difference is applied to a metallic conductor, the free electrons are forced to 'drift' through the lattice. But the motion of these electrons is impeded by their collisions with the lattice. This is how the resistive effect arises and, as the electrons transfer their energy to the lattice, the temperature of the metal will rise.

The magnitude of the current is given by the expression:

$$I = nqvA$$

where:

- n is the number of charge carriers per cubic metre (the carrier concentration)
- q is the charge on each charge carrier
- v is the drift velocity of the charge carriers
- A is the cross-sectional area of the conductor

Worked example

A tungsten filament lamp carries a current of 240 mA. The diameter of the filament is 0.024 mm and that of the copper connecting leads is 0.80 mm.

Calculate the drift velocity of the electrons in:

a the filament
b the connecting leads

You may take the number of free electrons per unit volume to be 8.0×10^{28} m^{-3} for copper and 4.0×10^{28} m^{-3} for tungsten.

Knowledge check 20

Three identical cells of emf 2.0 V and internal resistance 0.50 Ω are connected such that two are in parallel, and these are then connected in series with the third cell. Calculate: (a) the total emf and (b) the internal resistance of the combination.

Exam tip

For metal wires both A and n are constant. The charge q represents that of an electron (1.6×10^{-19} C). The larger the resistivity the lower the drift velocity becomes.

Answer

a $v = \dfrac{I}{nqA}$

$= \dfrac{0.240\,\text{A}}{(4.0 \times 10^{28}\,\text{m}^{-3}) \times (1.6 \times 10^{-19}\,\text{C}) \times \pi(0.012 \times 10^{-3})^2} = 0.083\,\text{m s}^{-1}$

b $v = \dfrac{0.240\,\text{A}}{(8.0 \times 10^{28}\,\text{m}^{-3}) \times (1.6 \times 10^{-19}\,\text{C}) \times \pi(0.40 \times 10^{-3})^2} = 3.7 \times 10^{-5}\,\text{m s}^{-1}$

The above calculation shows how slowly the electrons drift through the lattice. Although they can have random speeds averaging several hundred metres per second, it would take the electrons about 30 seconds to drift 1 millimetre in the copper leads, i.e. the drift velocity is very much smaller than the thermal speeds of the electrons.

Variations in resistivity

The equation $I = nqvA$ can be used to explain the variations in resistivity of different materials at different temperatures. For a given cross-section, the current depends on the carrier concentration and the drift velocity (in most cases q equals the electron charge $1.6 \times 10^{-19}\,\text{C}$).

The equation tells us that the higher the carrier concentration, the better the conductivity. For example, the value of n for copper is about $8.0 \times 10^{28}\,\text{m}^{-3}$ while that of the high-resistivity alloy nichrome is about $5.0 \times 10^{26}\,\text{m}^{-3}$.

Semiconducting materials such as silicon have much lower carrier concentrations — on the order of $10^{23}\,\text{m}^{-3}$ — and their resistivities are correspondingly higher than those of metals.

Resistivity also depends on temperature. The I–V curve for a tungsten filament lamp in Figure 34 illustrates the effect of temperature on the resistance of the tungsten filament. At higher voltages, when the filament is hot, the graph shows that the ratio of V to I rises and so the resistance increases.

Metals are said to have a *positive* **temperature coefficient of resistivity**. At higher temperatures, owing to increased thermal vibration of the lattice, there will be greater interaction between the free electrons and the lattice. This results in a reduced drift velocity and, consequently, a reduction in the current.

In contrast to metals, semiconductors have a *negative* temperature coefficient of resistivity. Thermistors are better conductors at higher temperatures — you may be familiar with the experiment demonstrating that a glass rod is insulating when cold but becomes conducting upon being strongly heated with a Bunsen flame. Although, like in metallic conductors, the increased lattice vibration will reduce the drift velocity v, the carrier concentration, n, increases markedly in thermistors as the temperature rises. In some cases, an overheated semiconductor will generate even more heat as the current rises, which further reduces the resistance, leading to an avalanche effect known as 'thermal runaway', and the eventual destruction of the component through melting.

Exam tip

When the value of the resistivity of a metal is quoted, the temperature must also be given. Many reference books give the values at 20°C (room temperature).

Knowledge check 21

Explain why the drift velocity of the charge carriers in a thermistor/semiconductor is much greater than that in a metallic conductor of the same dimensions and carrying the same current.

Knowledge check 22

State the difference between positive and negative coefficients of resistivity.

Electric circuits

In a light-dependent resistor (LDR) an increase in illumination will provide extra energy to release more conduction electrons. Thus the current increases and the LDR has a lower resistance. An LDR therefore has a negative temperature coefficient. As the number of conduction electrons increases very rapidly with illumination, the resistance decreases markedly.

Summary

After studying this section, you should be able to:
- define current, potential difference and resistance, and use the relationships between them to determine values of these quantities in series and parallel circuits
- use the equation $R = \rho l/A$ and explain the difference between resistance and resistivity
- state Ohm's law and distinguish between ohmic and non-ohmic conductors
- understand the concepts of electromotive force (emf) and internal resistance of a source of electrical energy, and describe how the internal resistance of a cell or battery is determined
- use the equation $I = nqvA$ to find the current or drift velocity in a conductor, and explain positive and negative coefficients of resistivity

Questions & Answers

The Edexcel examinations

This guide covers only the sections on mechanics and electrical circuits that are required for AS paper 1 and for part of the A-level paper 1 and paper 3.

The Edexcel AS examination consists of two papers, each containing multiple-choice, short-open, open-response, calculation and extended-writing questions. Both papers are of 1 hour and 30 minutes duration and have 80 marks. The examination is intended for students who have completed a 1-year course of study that is based on the core physics of the A-level specification.

The A-level examination consists of three papers. Paper 1 and paper 2 contain multiple-choice, short-open, open-response, calculation and extended-writing questions. Both papers are of 1 hour and 45 minutes duration and have 90 marks.

Paper 3 covers the general and practical principles of physics. It is of 2 hours and 30 minutes duration and has 120 marks. This paper covers all of the topics and includes synoptic questions as well as assessing the conceptual and theoretical understanding of experimental methods.

The AS papers consist of two sections, A and B. Section A will have 56–60 marks and section B 20–24 marks of the 80 available marks.
- Core Physics I covers the Mechanics and Electric circuits topics in Section A, while Section B will include a data-analysis question, possibly with an experimental context, and will draw on topics from the whole specification.
- Core Physics II covers Materials and waves and the Particle nature of light topics in Section A, while Section B will include a short article with questions drawn on topics from the whole specification.

All of the content in Core Physics I may be examined in paper 1 and paper 3 and the A-level examinations, and that in Core Physics II may appear in papers 2 and 3.

Note that both papers 1 and 2 at AS and A-level will also examine 'Working as a physicist'. Briefly, this means students:
- working scientifically, developing competence in manipulating quantities and their units, including making estimates
- experiencing a wide variety of practical work, developing practical and investigative skills by planning, carrying out and evaluating experiments and becoming knowledgeable of the ways in which scientific ideas are used
- developing the ability to communicate their knowledge and understanding of physics
- acquiring these skills through examples and applications from the entire course

A formulae sheet is provided with each test. Copies may be downloaded from the Edexcel website, or can be found at the end of past papers.

Command terms

Examiners use certain terms that require you to respond in a particular way. You must be able to distinguish between these terms and understand exactly what each requires you to do. A full list of command terms is given in the Edexcel specification. Some frequently used commands are shown below.

- **State** — a brief sentence giving the essential facts; no explanation is required (nor should you give one).
- **Define** — you can use a *word equation*; if you use *symbols*, you must state what each symbol represents.
- **List** — simply a series of words or terms, with no need to write sentences.
- **Outline** — a logical series of bullet points or phrases will suffice.
- **Describe** — for an experiment, a diagram is essential, then give the main points concisely (bullet points can be used).
- **Draw** — diagrams should be drawn in section, neatly and *fully labelled* with all measurements clearly shown, but don't waste time — remember it is not an art exam.
- **Sketch** — usually a graph, but graph paper is not necessary, although a grid is sometimes provided; axes must be labelled, including a scale if numerical data are given, the origin should be shown if appropriate, and the general shape of the expected line should be drawn.
- **Explain** — use correct physics terminology and principles; the depth of your answer should reflect the number of marks available.
- **Show that** — usually a value is given so that you can proceed with the next part; you should show all your working and give your answer to more significant figures than the value given (to prove that you have actually done the calculation).
- **Calculate** — show all your working and give *units* at every stage; the number of significant figures in your answer should reflect the given data, but you should keep each stage in your calculator to prevent rounding errors.
- **Determine** — you will probably have to extract some data, often from a graph, in order to perform a calculation.
- **Estimate** — a calculation in which you have to make a sensible assumption, possibly about the value of one of the quantities — think, does this give a reasonable answer?
- **Suggest** — there is often no single correct answer; credit is given for sensible reasoning based on correct physics.
- **Discuss** — you need to sustain an argument, giving evidence for and against, based on your knowledge of physics and possibly using appropriate data to justify your answer.

You should pay particular attention to diagrams, sketching graphs and calculations. Students often lose marks by failing to label diagrams properly, by not giving essential numerical data on sketch graphs and, in calculations, by not showing all the working or by omitting the units.

Questions & Answers

About this section

The following two tests are made up of questions similar in style and content to the AS and A-level examinations. The first test is in the style of an AS paper 1, but section B will be restricted to the topics in this guide.

You may like to attempt a complete paper in the allotted time and then check your answers, or maybe do the multiple-choice section and selected questions to fit your revision plan. It is worth noting that there are 80 marks available for the 90-minute test, so this should help in determining how long you should spend on a particular question. You should therefore be looking at about 10 minutes for the multiple-choice section and just over a minute a mark on the others.

The second test is in the style of the A-level paper 1, but again is restricted to the topics covered in this guide.

You should also be aware that during the examination you must write your answers directly onto the paper. This will not be possible for the tests in this book, but the style and content are the same as the examination scripts in every other respect. It may be that diagrams and graphs that would normally be added to on the paper, have to be copied and redrawn. If you are doing a timed practice test, you should add an extra few minutes to allow for this.

The answers should not be treated as model answers because they represent the bare minimum necessary to gain the marks. In some instances, the difference between an A-grade response and a C-grade response is suggested. This is not possible for the multiple-choice section, and many of the shorter questions that do not require extended writing.

Ticks (✓) are included in the answers to indicate where the examiner has awarded a mark. Half marks are not given.

AS Test Paper 1

Time allowed: 1 hour 30 minutes. Answer ALL the questions.

Section A

For Questions 1–8 select one answer from A to D.

Question 1

A car accelerates uniformly from $10\,\text{m s}^{-1}$ to $16\,\text{m s}^{-1}$ in a time of 1 minute 40 seconds. The value of the acceleration during this time is:

A $0.01\,\text{m s}^{-2}$ B $0.06\,\text{m s}^{-2}$ C $0.10\,\text{m s}^{-2}$ D $0.16\,\text{m s}^{-2}$ (1 mark)

Question 2

The distance travelled by the car in Question 1 during this 1 minute 40 seconds is:

A $0.50\,\text{km}$ B $1.0\,\text{km}$ C $1.3\,\text{km}$ D $1.6\,\text{km}$ (1 mark)

Question 3

Electrical power can be given by the expression:

A $\dfrac{QI}{t}$ B $\dfrac{QV}{t}$ C $\dfrac{RI}{t}$ D $\dfrac{VI}{t}$ (1 mark)

Question 4

The resistance of $5.00\,\text{m}$ of wire of diameter $1.0\,\text{mm}$ and resistivity $1.1 \times 10^{-6}\,\Omega\,\text{m}$ is:

A $1.8\,\Omega$ B $5.5\,\Omega$ C $7.0\,\Omega$ D $22\,\Omega$ (1 mark)

Question 5

Which of the following is **not** a vector quantity?

A acceleration   B displacement   C potential energy   D velocity   (1 mark)

Question 6

A golf ball leaves the club with a velocity of $30\,\text{m s}^{-1}$ at an angle of $30°$ to the ground. The horizontal component of the velocity is:

A $10\,\text{m s}^{-1}$ B $14\,\text{m s}^{-1}$ C $20\,\text{m s}^{-1}$ D $26\,\text{m s}^{-1}$ (1 mark)

Ⓔ Questions 7 and 8 relate to the resistor network shown below:

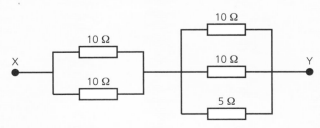

Questions & Answers

Question 7
The total resistance of the network is

A 7.5Ω   B 12.5Ω   C 22.5Ω   D 45Ω   (1 mark)

Question 8
When a potential difference of 12V is applied across X and Y, the current in the 5.0Ω resistor is:

A 0.16A   B 0.80A   C 1.6A   D 4.8A   (1 mark)

Total: 8 marks

Answers to Questions 1–8

(1) B

   ⓔ Use $v = u + at$. Convert the time to 100s. $a = 0.06\,\text{m s}^{-2}$

(2) C

   ⓔ Use s = average velocity × time, or $v^2 = u^2 + 2as$. $s = 1300\,\text{m}$

(3) B

   ⓔ Use $P = IV$ and $I = Q/t$, or $P = W/t$ and $W = QV$

(4) C

$$R = \frac{\rho l}{A} = \frac{(1.1 \times 10^{-6}\,\Omega\,\text{m}) \times 5.00\,\text{m}}{\pi (0.5 \times 10^{-3})^2\,\text{m}^2} = 7.0\,\Omega$$

   ⓔ A common error is to use the diameter rather than the radius in finding the area.

(5) C

   ⓔ No direction can be assigned to energy and so it is a scalar quantity.

(6) D

   ⓔ Horizontal component = $30 \cos 30\,\text{m s}^{-1} = 26\,\text{m s}^{-1}$

(7) A

   ⓔ This network consists of two parallel circuits connected in series. The resistance of the parallel pair on the left-hand side is 5.0Ω, and the resistance of the three-resistor circuit on the right-hand side is 2.5Ω. The total resistance is therefore 7.5Ω.

(8) B

   ⓔ There will be 8.0V across the first pair of resistors and 4.0V across the other three (including the 5Ω resistor). So:

$$I = \frac{4.0\,\text{V}}{5.0\,\Omega} = 0.80\,\text{A}$$

Question 9

An electric motor is used to raise a load through a measured distance. A student uses the arrangement to investigate the power output of the motor. Her results are shown in Table 1.

Load/kg	Height/m	Time/s	Work done/J	Power/W
0.200	1.20	0.42		
0.400	1.20	0.78		

Table 1

(a) Copy and complete the table to show the work done in raising the load and the output power for both loads. (2 marks)

ⓔ Ensure that the mass is multiplied by $9.8\,\mathrm{m\,s^{-2}}$ when calculating work done.

(b) The motor is connected to a 12 V supply and the average current that flows through it when raising the 0.400 kg load is 0.75 A. Calculate the efficiency of the motor when raising the 0.400 kg load. (2 marks)

Total: 4 marks

Student answer

(a) work done = force × distance

0.200 kg load: work done = $0.200\,\mathrm{kg} \times 9.8\,\mathrm{m\,s^{-2}} \times 1.20\,\mathrm{m} = 2.4\,\mathrm{J}$

0.400 kg load: work done = $0.400\,\mathrm{kg} \times 9.8\,\mathrm{m\,s^{-2}} \times 1.20\,\mathrm{m} = 4.7\,\mathrm{J}$ ✓

power = $\dfrac{\text{work done}}{\text{time}}$

0.200 kg load: power = $\dfrac{2.4\,\mathrm{J}}{0.42\,\mathrm{s}} = 5.7\,\mathrm{W}$

0.400 kg load: power = $\dfrac{4.7\,\mathrm{J}}{0.78\,\mathrm{s}} = 6.0\,\mathrm{W}$ ✓

ⓔ Students who forget to include g in the calculation can still gain the second mark (answers 0.57 W and 0.61 W).

(b) electrical power supplied to the motor = $V \times I$

$= 12\,\mathrm{V} \times 0.75\,\mathrm{A} = 9.0\,\mathrm{W}$ ✓

efficiency = $\dfrac{\text{power output}}{\text{power input}} \times 100\%$

$= \dfrac{6.0\,\mathrm{W}}{9.0\,\mathrm{W}} \times 100\% = 67\%$ ✓

Questions & Answers

Question 10

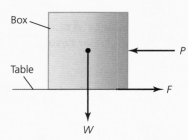

A box is pushed along a table at constant speed. The box is in *equilibrium*.

(a) Explain the meaning of 'equilibrium'. (1 mark)

(b) Copy the diagram and add the fourth force needed for the box to be in equilibrium. (1 mark)

(c) State Newton's third law of motion. (2 marks)

(d) Copy and complete Table 2. (4 marks)

Force	Description of force	Body upon which the Newton's third law pair acts	Type of force	Direction of the Newton's third law pair force
W				
F				

Table 2

Total: 8 marks

ⓔ This question tests the application of Newton's first and third laws. Part (a) requires you to consider the forces acting on a single body, and the table in part (d) tests your understanding of Newton's third law pairs. A grade A student will be aware that, in addition to being equal and opposite in direction, the pairs must be of the same type.

Student answer

(a) Equilibrium is the condition of a body when the resultant force (in any two directions) is zero. ✓

(b) Arrow drawn from the surface, upward and along the same line as the weight (or somewhere between the line of the weight and the left-hand edge of the block). ✓

(c) If body A exerts a force on body B, then body B will exert an equal ✓ and opposite ✓ force on body A.

(d)

Force	Description of force	Body upon which the Newton's third law pair acts	Type of force	Direction of the Newton's third law pair force
W	Weight	The Earth	Gravitational (*not* just gravity)	Up(ward)
F	Friction	Table	Contact	Right to left (opposite from F)
	✓	✓	✓	✓

ⓔ 1 mark for each correct column.

Question 11

On the diagram below, graph X shows how the potential difference across the terminals of a cell depends on the current in the cell. Graph Y is the voltage–current characteristic for a filament lamp.

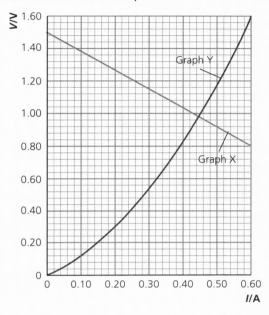

e This question examines the relationship between the emf, internal resistance and terminal potential difference for a cell (see the equation on p. 44) and the variable resistance of a filament lamp at different temperatures. Note that the axes of the graph are the opposite of those normally used for the characteristics of a lamp.

(a) As the current increases, what can be deduced from the graphs about:
   (i) the internal resistance of the cell?
   (ii) the resistance of the filament lamp? (2 marks)

(b) Use graph X to determine:
   (i) the emf of the cell
   (ii) the internal resistance of the cell (2 marks)

e The question requires that the information is to be determined from the graph. Students who try to calculate the values from other data will gain no marks.

(c) When the lamp is connected to the cell, what is:
   (i) the current in the lamp?
   (ii) the resistance of the lamp?
   (iii) the power developed in the lamp? (3 marks)

(d) Draw a circuit diagram of the circuit you would set up to obtain the data for graph Y, using two cells connected in series, which are identical to the one in graph X, for the power supply. (3 marks)

Questions & Answers

e The circuit must include the appropriate meters, correctly connected, and some means of varying the current or potential difference.

(e) Explain:
  (i) why you would need two cells in series, rather than a single cell, to achieve the results shown by graph Y
  (ii) how you would obtain the data (3 marks)

Total: 13 marks

e You will need to consider the effect of the internal resistance of the cell on the maximum terminal potential difference.

Student answer

(a) (i) The internal resistance is constant. ✓

  (ii) The resistance of the lamp rises as the voltage or current increases. ✓

(b) Graph X represents the relationship $V = \varepsilon - Ir$.

  (i) The emf ε is the intercept on the V-axis, i.e. 1.52 V. ✓

  (ii) The internal resistance r is the negative of the gradient.
  $$r = \frac{(0.8 - 1.52)\,V}{0.60\,A} = 1.2\,\Omega \checkmark$$

(c) When the lamp is connected to the cell, the values of V and I will correspond to the coordinates of the point of intersection between graph X and graph Y.

  (i) $I = 0.45\,A$ ✓

  (ii) $V = 0.98\,V$, so $R = \dfrac{0.98\,V}{0.45\,A} = 2.2\,\Omega$ ✓

  (iii) $P = VI = 0.44\,W$ ✓

(d) The diagram should show:

  a battery of two cells ✓

  a means of varying the voltage (e.g. a potential divider) ✓

  an ammeter in series and a voltmeter in parallel with the lamp ✓

(e) (i) With a single cell connected to the lamp, the maximum terminal potential difference would be less than 1.6 V. So, to cover the full range (up to about 1.6 V), a second cell is needed. ✓

  (ii) Adjust the potentiometer or variable resistor to give a range of voltages (e.g. 0 to 1.60 V at 0.20 V intervals) ✓ and read off the corresponding currents from the ammeter ✓.

Question 12

The graph shows the variation of velocity with time for a ball released from rest and allowed to bounce off the floor.

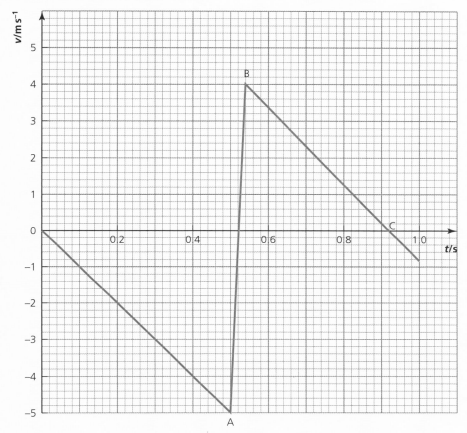

(i) Determine the gradient of the region 0A of the graph. What quantity does this represent? (2 marks)

(ii) State what is happening to the ball during the interval between A and B. (1 mark)

(iii) Calculate the displacement of the ball at point C. (3 marks)

Total: 6 marks

ⓔ This question tests the student's ability to analyse a velocity–time graph. A C-grade student may gain some marks by finding a displacement using the area under the graph, but an A-grade student will recognise that displacement is a vector quantity. As downward velocities are assigned negative values, downward displacement and accelerations are also negative. In this case, for full marks, the acceleration must be $-10\,\text{m s}^{-2}$, and the displacement at point C = $-0.45\,\text{m}$, i.e. $0.45\,\text{m}$ below the point of release.

Questions & Answers

> **Student answer**
>
> (i) gradient = $\dfrac{-5\,\text{m s}^{-1}}{0.5\,\text{s}} = -10\,\text{m s}^{-2}$ ✓

ⓔ To gain this mark both the unit and the negative sign are needed.

> The gradient represents acceleration (due to gravity). ✓
>
> (ii) AB represents the time that the ball is in contact with the ground/changing direction/bouncing. ✓
>
> (iii) Displacement is equal to the area under a velocity–time graph. ✓
>
> $= \dfrac{1}{2}(0.50\,\text{s}) \times (-5.0\,\text{m s}^{-1}) + \dfrac{1}{2}(0.40\,\text{s}) \times (4.0\,\text{m s}^{-1})$ ✓ (correct values regardless of sign)
>
> $= -1.25\,\text{m} + 0.80\,\text{m} = -0.45\,\text{m}$ ✓

Question 13

(a) State the principle of conservation of linear momentum. (2 marks)

A student investigating the law uses an air track with two gliders as shown in the diagram:

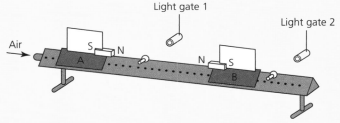

The student pushes glider A towards glider B, which is initially stationary. As glider A approaches glider B, the magnetic poles repel so that B is pushed forward and A moves backward, in the opposite direction from its original motion. The student measures the time for the interrupter card to cross the light gates before and after the interaction. Her results are given in Table 3 below:

Mass of A = 200 g   Mass of B = 400 g   Length of card = 20.0 cm

Glider	t/s	v/m s^{-1}	p/kg m s^{-1}
A (before)	0.221	0.905	0.181
A (after)	0.659		
B (after)	0.331		

Table 3

(b) Complete the table to show the velocity and momentum of the gliders after the collision. (2 marks)

ⓔ Remember that momentum and velocity are vectors. The direction must be indicated.

(c) Show whether or not these results confirm that momentum is conserved in the interaction. (2 marks)

During the collision, the momentum of each glider changes. Newton's second law of motion can be expressed in the form *The resultant force acting upon a body equals the rate of change in momentum of the body*.

(d) If a body of mass m, changes velocity from u to v, in a time Δt, show that Newton's second law may be written as:

$F = ma$

where F is the resultant force and a is the acceleration of the mass during the collision. (2 marks)

The duration of the collision in part (a) is 78 ms.

(e) Use the data from your table to calculate the average force exerted on glider A and on glider B during the collision. (2 marks)

(f) Explain how the values of the forces on each glider demonstrate the validity of Newton's third law. (2 marks)

Total: 12 marks

> **Student answer**
>
> **(a)** In any system of interacting bodies, the total momentum is conserved, provided that no resultant ✓ external force acts on the system. ✓
>
> **(b)** After the collision:
>
> velocity of A = $\dfrac{-0.20\,\text{m}}{0.659\,\text{s}}$ = $-0.303\,\text{m s}^{-1}$
>
> velocity of B = $\dfrac{0.20\,\text{m}}{0.331\,\text{s}}$ = $0.604\,\text{m s}^{-1}$ ✓
>
> momentum of A = $0.20\,\text{kg} \times (-0.303\,\text{m s}^{-1})$ = $-0.061\,\text{kg m s}^{-1}$
>
> momentum of B = $0.40\,\text{kg} \times 0.604\,\text{m s}^{-1}$ = $0.242\,\text{kg m s}^{-1}$ ✓

ℯ A typical error made by a grade-C student is to omit the negative signs for the velocity and momentum of A after the collision; this would lead to one of the marks being lost.

> **(c)** momentum before the collision = $0.181\,\text{kg m s}^{-1}$
>
> total momentum after the collision = $-0.061\,\text{kg m s}^{-1} + 0.242\,\text{kg m s}^{-1}$ ✓
>
> $= 0.181\,\text{kg m s}^{-1}$
>
> So momentum is conserved. ✓
>
> **(d)** $F = \dfrac{\text{change in momentum}}{\Delta t} = \dfrac{mv - mu}{\Delta t} = m\dfrac{v - u}{\Delta t} = ma$ ✓✓
>
> **(e)** force on A = $\dfrac{0.181\,\text{kg m s}^{-1} - (-0.061\,\text{kg m s}^{-1})}{0.078\,\text{s}}$ = $3.1\,\text{N}$ ✓
>
> force on B = $\dfrac{0\,\text{kg m s}^{-1} - 0.242\,\text{kg m s}^{-1}}{0.078\,\text{s}}$ = $-3.1\,\text{N}$ ✓
>
> **(f)** The forces are equal but in the opposite direction ✓. This is in agreement with Newton's third law ✓.

Questions & Answers

Question 14

The diagram shows the forces acting on a suitcase held at rest by a vertical upward force, F, at the handle. The case and its contents have a total mass of 8.0 kg and the centre of gravity is labelled G.

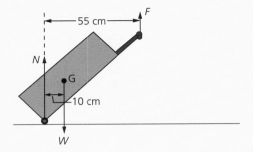

(a) Explain what is meant by the *centre of gravity* of the case. (1 mark)

(b) Use Newton's first law to write an equation relating the size of the forces N, W and F. (1 mark)

(c) Use the principle of moments to find the magnitude of force F. (2 marks)

(d) Explain how the value of F would be affected if the centre of gravity was lower down in the case. (3 marks)

Total: 7 marks

Total for Section A: 58 marks

Student answer

(a) The centre of gravity is the point where all of the weight of an object can be said to act. ✓

ⓔ Many students lose this mark by omitting the word 'point'.

(b) As the case is in a state of equilibrium, the resultant force acting upon it must be zero (Newton's first law), so $F + N = W$. ✓

(c) Taking moments about the pivot (the axle of the wheel):

clockwise moment = $W \times 0.10\,m = 8.0\,kg \times 9.8\,ms^{-2} \times 0.10\,m = 7.8\,Nm$

= anticlockwise moment = $F \times 0.55\,m$ ✓

$F = 14\,N$ ✓

(d) If G were lower, the perpendicular distance of the line of action of W would be less, resulting in a lower value for the clockwise moment. ✓ For equilibrium the anticlockwise moment must also decrease. ✓ Since the line of action of F is at the same perpendicular distance from the wheel as before, the value of F must reduce. ✓

Section B

Question 15

Two students are given a metre rule, a marble and a stopwatch and asked to find a value of the acceleration due to gravity. One student measures a distance of 2.00 m from the floor and makes a mark on the wall. She then drops the marble from the height of the mark and the other student takes the time for the marble to fall to the floor.

Table 4 shows the times taken for five repeats of the exercise.

t_1/s	t_2/s	t_3/s	t_4/s	t_5/s	Mean time/s
0.68	0.64	0.70	0.54	0.66	0.67

Table 4

(a) Explain why the students concluded that the mean value of the time should be 0.67 s. (1 mark)

(b) Calculate the percentage uncertainty of the students' measurements. You may assume that the height has an uncertainty of ±0.5 cm. (2 marks)

(c) Use the students' readings to calculate a value for the acceleration due to gravity. (2 marks)

(d) The accepted value of g is 9.8 m s^{-2}. Show that this is outside of the range given by the uncertainty of the measurements, and suggest a reason for this discrepancy. (2 marks)

Total: 7 marks

> **Student answer**
>
> (a) The value of t_4 is very much less than the rest. It is probably a mistake by the timer and so it is ignored when taking the average value. ✓
>
> (b) % uncertainty in height = $\dfrac{0.005\,\text{m}}{2.00\,\text{m}} \times 100\% = 0.25\%$ ✓
>
> uncertainty in time = ± 0.03 s from mean of 0.67 s
>
> % uncertainty in time = $\dfrac{0.03\,\text{s}}{0.67\,\text{s}} \times 100\% = 4.5\%$ ✓
>
> (c) Using $s = ut + \frac{1}{2}at^2$:
>
> $2.00\,\text{m} = 0 + \frac{1}{2}g(0.67\,\text{s})^2$ ✓ $g = 8.9\,\text{m s}^{-2}$ ✓
>
> (d) 8.9 m s^{-2} ± 4.5% = 8.9 m s^{-2} ± 0.4 m s^{-2} ✓
>
> The value of 9.8 m s^{-2} is outside the student's range, probably due to reaction times of the student manually pressing the stopwatch. ✓

Questions & Answers

Question 16

A student is given a piece of wire about 80 cm long and half a millimetre in diameter and asked to determine the material of the wire by measuring its resistivity. He was also given the following resistivities (Table 5) of some metals:

Metal	Resistivity (at 20°C)/Ω m
Aluminium	2.8×10^{-8}
Copper	1.7×10^{-8}
Constantan	4.9×10^{-7}
Nichrome	1.1×10^{-6}

Table 5

(a) What is meant by resistivity? (1 mark)

The student used a micrometer to measure the diameter at four different positions along the wire. His results are shown below:

d/mm   0.32   0.31   0.31   0.32

(b) (i) Why was a micrometer a suitable instrument for measuring the diameter of the wire?
  (ii) Why were measurements taken at four different places?
  (iii) Calculate the area of cross-section of the wire. (3 marks)

The student set up the circuit shown below in order to measure the resistance of the wire for a range of lengths.

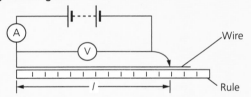

His teacher suggested that he should use a variable resistor in the circuit to ensure that the current didn't exceed 500 mA.

(c) Why is such a small current beneficial? (1 mark)

The student presented his results as shown in Table 6:

l/m	V/V	I/mA	R/Ω
0.1	0.37	220	1.7
0.2	0.68	220	3.1
0.3	0.98	220	4.5
0.4	1.30	220	5.9
0.5	1.61	220	7.3
0.6	1.91	220	8.7

Table 6

(d) State the error that the student has incurred in the presentation of these readings. (1 mark)

(e) Plot a graph of resistance against length. (4 marks)

(f) Use your graph and your value for the area of cross-section of the wire to determine which material is used for the wire. (3 marks)

It is suggested that the contact resistance at each end of the wire could have caused a systematic error in the values of R.

(g) Explain what is meant by a systematic error, and describe how this is evident on the graph.
(2 marks)

Total: 15 marks

Total for Section B: 22 marks

Total for paper: 80 marks

> **Student answer**
>
> **(a)** Resistivity is the property of a material to resist the flow of charge. It is defined by the equation $\rho = RA/l$. ✓
>
> **(b) (i)** A micrometer has a scale that reads to divisions of 0.01 mm and so is suitable for wires of diameter of the order of 0.5 mm. ✓
>
> **(ii)** The diameter is measured at different places to check that it is of uniform diameter. ✓
>
> **(iii)** $A = \dfrac{\pi d^2}{4} = 7.8 \times 10^{-8} \, m^2$ ✓
>
> **(c)** Large currents make the wire heat up, causing an increase in the resistivity of the wire. ✓
>
> **(d)** The values of l are written to only one significant figure, although the readings on the metre rule are to ±1 mm (0.001 m). ✓
>
> **(e)** Axes correctly labelled ✓, scale ✓, points drawn ✓, line of best fit ✓.
>
>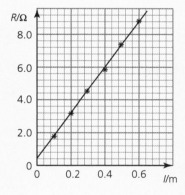

ⓔ Many students lose this mark by forcing the line through the origin.

> **(f)** The gradient of the graph is $14 \, \Omega \, m^{-1}$ ✓
>
> $\rho = \dfrac{RA}{l}$ = gradient × A = $14 \, \Omega \, m^{-1} \times 7.8 \times 10^{-8} \, m^2 = 1.1 \times 10^{-6} \, \Omega \, m$ ✓
>
> The wire is made from nichrome. ✓
>
> **(g)** A systematic error is one that is the same for every reading, for example a zero error on a rule (or micrometer). ✓
>
> This is shown by the intercept on the R axis, showing that every value of the resistance is greater than the resistance of the wire by the same amount (about $0.4 \, \Omega$) for every reading. ✓

A-level Test Paper 1

Time allowed: 1 hour 45 minutes. Answer ALL the questions.

For Questions 1–8 select one answer from A to D.

Question 1

The watt can be expressed in SI base units as:

A $kg\,m\,s^{-3}$ B $kg\,m^2\,s^{-1}$ C $kg\,m^2\,s^{-2}$ D $kg\,m^2\,s^{-3}$ (1 mark)

ⓔ Questions 2 and 3 relate to the potential divider circuit shown below.

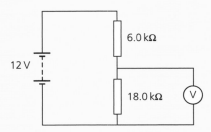

Question 2

Assuming that the voltmeter has an infinite resistance, it will read:

A 3.0 V   B 4.0 V   C 6.0 V   D 9.0 V   (1 mark)

Question 3

The voltmeter is replaced by another one that has a resistance of 36 Ω. The reading will now be:

A 4.0 V   B 6.0 V   C 8.0 V   D 9.0 V   (1 mark)

Question 4

A high-jumper has an initial upward velocity of $6.0\,m\,s^{-1}$. The maximum height reached by her centre of gravity will be about:

A 1.6 m   B 1.8 m   C 2.0 m   D 2.2 m   (1 mark)

Question 5

The kinetic energy of an object moving at a velocity v and having momentum p is:

A $\tfrac{1}{2}pv$ B pv C $\tfrac{1}{2}p^2v$ D $\tfrac{1}{2}pv^2$ (1 mark)

Question 6

An electric kettle has a heating element of resistance 19 Ω and is connected to a 230 V supply. The energy transferred to the water in the kettle when it is switched on for 3.0 minutes is:

A 2.2 kJ   B 8.4 kJ   C 500 kJ   D 790 kJ   (1 mark)

A-level Test Paper 1

Question 7

A pool cue of length 1.60 m and weight 56 N is balanced on a pivot placed at its mid-point by hanging a mass of weight 21 N from its tip, as shown in the diagram.

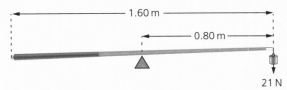

The distance of the centre of gravity from the tip of the cue is:

A 0.30 m   **B** 0.50 m   **C** 0.80 m   **D** 1.10 m   (1 mark)

Question 8

Select the graph that best describes how the resistance of a NTC thermistor varies with the temperature θ.

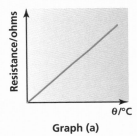
Graph (a)

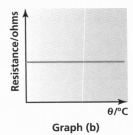
Graph (b)

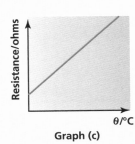
Graph (c)

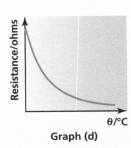
Graph (d)

A graph (a)   **B** graph (b)   **C** graph (c)   **D** graph (d)   (1 mark)

In questions 9 and 10, which of the following graphs best represent the quantities described when they are plotted on the y- and x-axes? Each graph may be used once, more than once or not at all.

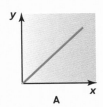
A

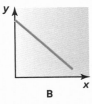
B

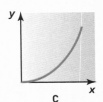
C

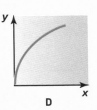
D

Mechanics • Electric circuits

Questions & Answers

Question 9
y-axis: kinetic energy of a body accelerating from rest
x-axis: velocity of the body
(1 mark)

Question 10
y-axis: velocity of a car that is accelerating uniformly
x-axis: time after the application of the brakes
(1 mark)

Answers to Questions 1–10

(1) D

ⓔ power = $\dfrac{\text{force} \times \text{distance}}{\text{time}} = \dfrac{\text{kg m s}^{-2} \times \text{m}}{\text{s}} = \text{kg m}^2\text{s}^{-3}$

(2) D

ⓔ $V_{out} = \dfrac{R_2}{R_1 + R_2} \times 12\text{V} = \dfrac{18\,\text{k}\Omega}{18\,\text{k}\Omega + 6.0\,\text{k}\Omega} \times 12\text{V} = 9.0\text{V}$

(3) C

ⓔ resistance of the 18 kΩ resistor and the 36 kΩ voltmeter in parallel = 12 kΩ

$V_{out} = \dfrac{R_2}{R_1 + R_2} \times 12\text{V} = \dfrac{12\,\text{k}\Omega}{12\,\text{k}\Omega + 6.0\,\text{k}\Omega} \times 12\text{V} = 8.0\text{V}$

(4) B

ⓔ Use the equation $v^2 = u^2 + 2as$. The velocity at the maximum height, *v*, is zero.
$0 = (6.0\,\text{m s}^{-1})^2 - 2 \times 9.8\,\text{m s}^{-2} \times s \quad s = 1.8\,\text{m}$

(5) A

ⓔ $p = mv \quad E_k = \tfrac{1}{2}mv^2 = \tfrac{1}{2}(mv)v = \tfrac{1}{2}pv$

(6) C

ⓔ $E = VI\Delta t = 230\text{V} \times \dfrac{230\text{V}}{19\,\Omega} \times 180\,\text{s} = 500\,\text{kJ}$

(7) D

ⓔ Using the principle of moments: 56 N × *x* = 21 N × 0.80 m; *x* = distance of the centre of gravity from the pivot = 0.30 m; distance to the tip = 1.10 m

(8) D

ⓔ A thermistor is a semiconducting device and so has a negative coefficient of resistance; as the temperature rises, the carrier concentration increases and so the resistance falls.

(9) C

ⓔ Kinetic energy = $\tfrac{1}{2}mv^2$, so *y* is proportional to x^2 (parabolic shape).

(10) B

Acceleration is represented by the gradient of a velocity–time graph. A uniform, negative gradient represents constant negative acceleration.

A-level Test Paper 1

Question 11

A 1.0 kΩ carbon resistor has a power rating of 100 mW.

(a) Calculate the maximum current that can safely pass through the resistor. (1 mark)

(b) A student designs a circuit in which the resistor is connected between the terminals of a 12 V supply, but finds that the resistor burns out.
Explain why the resistor failed. (1 mark)

(c) His teacher recommends that he replace the resistor with four 1.0 kΩ resistors connected as shown below:

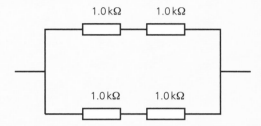

Show that the total resistance of the network is 1.0 kΩ. (1 mark)

(d) Explain why the resistors in this network will not burn out when connected to the 12 V supply. (2 marks)

Total: 5 marks

ⓔ This question requires the use of the basic equations for electrical power and parallel circuits.

Student answer

(a) Using $P = I^2R$:

$$I^2_{max} = \frac{P_{max}}{R} = \frac{100 \times 10^{-3}\,W}{1.0 \times 10^3\,\Omega} = 1 \times 10^{-4}\,A^2$$

so

$I_{max} = 10^{-2}\,A = 10\,mA$ ✓

(b) When a voltage of 12 V is applied across a 1.0 kΩ resistor, there is a current of 12 mA in the resistor. This is greater than the maximum current that the resistor can take, so it will burn out. ✓

(c) The network is the same as two 2.0 kΩ resistors in parallel:

$$\frac{1}{R} = \frac{1}{2.0 \times 10^3\,\Omega} + \frac{1}{2.0 \times 10^3\,\Omega} = \frac{1}{1 \times 10^{-3}\,\Omega^{-1}} \Rightarrow R = 1.0\,k\Omega\ \checkmark$$

(d) The current in each arm of the network is:

$$\frac{12\,V}{2.0\,k\Omega} = 6.0\,mA\ \checkmark$$

The current in each resistor is less than 10 mA, so they will not overheat. ✓

Questions & Answers

ⓔ A C-grade student may gain the marks for calculating the values of the current in parts (b) and (d) but lose the final mark if the values are not compared with the maximum value in part (a).

Question 12

A teacher places a coin on the edge of a bench and another, identical, coin on the end of a metre rule, as shown in the diagram. The rule is moved sharply across the surface so that the coin on the rule is left behind and falls directly to the ground and the one on the bench is projected horizontally.

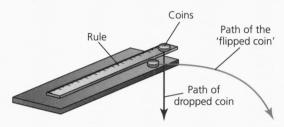

ⓔ This question depends on the understanding that the vertical and horizontal motions of a free-falling body can be treated independently.

(a) Explain why both coins hit the floor at the same instant. (2 marks)

(b) The height of the bench is 0.82 m and the projected coin leaves the bench with a velocity of $3.2\,m\,s^{-1}$.

　(i) Show that the time taken for the coins to reach the floor is about 0.4 s. (2 marks)

ⓔ This is a 'show that' question, so your calculated value should be to better than one significant figure.

　(ii) Calculate the horizontal distance travelled by the projected coin when it strikes the floor. (1 mark)

(c) The teacher repeats the demonstration but replaces the coin on the bench with a larger coin with a greater mass than the coin on the rule. Assuming the coin leaves the bench with a velocity of $3.2\,m\,s^{-1}$, as before, compare its trajectory with that of the coin projected in the first demonstration. Explain your answer. (3 marks)

Total: 8 marks

ⓔ This part requires an explanation of Galileo's hypothesis; i.e. near the surface of the Earth all objects accelerate at the same rate (ignoring air resistance). The ratio F/m is constant in a uniform gravitational field.

A-level Test Paper 1

Student answer

(a) Horizontal and vertical motions are independent of each other. ✓

Both coins accelerate downwards at the same rate ($9.8\,\mathrm{m\,s^{-2}}$). ✓

(b) (i) For the vertical motion use $s = ut + \frac{1}{2}at^2$ (or a combination of the two other equations).

$0.82\,\mathrm{m} = 0 + \frac{1}{2} \times 9.8\,\mathrm{m\,s^{-2}} \times t^2$ ✓ $\Rightarrow t = 0.41\,\mathrm{s}$ ✓

(ii) horizontal displacement = velocity × time = $3.2\,\mathrm{m\,s^{-1}} \times 0.41\,\mathrm{s} = 1.3\,\mathrm{m}$ ✓

(c) If air resistance is negligible the trajectory will be the same in both cases. ✓

The acceleration due to gravity is the same for all masses. ✓

(There is no component of the weight in the horizontal direction) so the horizontal velocity remains constant. ✓

Question 13

Under steady light conditions, the emf of a photovoltaic cell is measured by connecting a voltmeter of very high resistance directly across the cell's terminals. The emf of a particular cell is found to be 1.60 V.

When a 100 Ω resistor is connected between the terminals of this cell, the potential difference across it is 1.54 V.

(a) Show that the internal resistance of the cell is about 4 Ω. (2 marks)

ⓔ Remember that, since this is a 'show that' question, the result must be given to one significant figure more than the value stated in the question.

A student wants to investigate how the maximum power output of the cell depends on the intensity of the light falling on it. He has read that the maximum output power will be achieved when the external resistance is equal to the internal resistance, so he replaces the 100 Ω resistor with one having the same value as the internal resistance of the cell.

(b) What will be the value of the voltage across the resistor if the light conditions are unchanged? (2 marks)

(c) Describe how the student would perform the investigation. You should state what additional equipment is required, draw a circuit diagram and describe the measurements that need to be taken. (6 marks)

ⓔ Descriptions of standard experiments are often given marks for coherent structure and no lines of reasoning. Answers need to be grammatically correct with few spelling mistakes, and must contain the appropriate technical and scientific terminology. A series of phrases or bullet points is acceptable.

Total: 10 marks

Questions & Answers

Student answer

(a) $I = \dfrac{V}{R} = \dfrac{1.54\,\text{V}}{100\,\Omega} = 0.0154\,\text{A}$ ✓

From $\varepsilon = V + Ir$ we have:

$r = \dfrac{\varepsilon - V}{I} = \dfrac{1.60\,\text{V} - 1.54\,\text{V}}{0.0154\,\text{A}} = 3.9\,\Omega$ ✓

(b) When $R = r = 4\,\Omega$:

$I = \dfrac{1.60\,\text{V}}{4.0\,\Omega + 4.0\,\Omega} = 0.20\,\text{A}$ ✓ (or 0.21 A if $R = 3.9\,\Omega$ used)

so voltage across the resistor is $0.20\,\text{A} \times 4.0\,\Omega = 0.80\,\text{V}$ ✓

e When the external resistance equals the internal resistance of the cell, the terminal voltage will equal half the value of the emf, as the same work has to be done in passing the current through the cell as in the external circuit.

(c) Additional equipment: ammeter and light meter. ✓

Circuit diagram should show the ammeter in series and voltmeter in parallel with the resistor. ✓

For a range of light intensities ✓ determine the internal resistance for each intensity. ✓ Using resistors having the same value as the internal resistance, measure the corresponding values of I and V ✓ so that the power can be calculated using $P = IV$. ✓

(d) It must be assumed that the internal resistance of the cell remains the same for all light conditions. ✓

Question 14

Fletcher's trolley was a common piece of equipment used in schools several decades ago to investigate the relationship $F = ma$. The trolley had five identical, cylindrical masses that were slotted into the body of the trolley. The masses were removed one at a time and added to the load, so that a range of forces could be applied to the trolley.

A-level Test Paper 1

A horizontal vibrating strip was fitted with a marker so that a trace was drawn on a card as the trolley moved along the track.

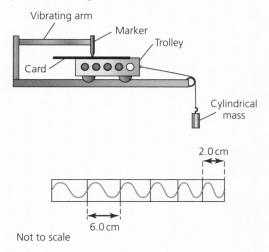

Not to scale

(a) Why is it necessary to increase the accelerating force by adding masses from the trolley to the load? (1 mark)

(b) For the sample card shown, the frequency of the vibrations of the strip is 5.0 Hz — i.e. each complete oscillation takes a time of 0.20 s.
Use the distances given on the diagram (which is not to scale) to calculate the average velocity of the trolley during the first marked oscillation and during the fifth cycle. Hence determine the acceleration of the trolley. (3 marks)

(c) In schools nowadays, there are several variations of this experiment. Describe an alternative method of determining the acceleration of the trolley. List any additional apparatus that would be required, and what readings need to be taken. (3 marks)

ⓔ A standard laboratory method to test the relationship between acceleration and applied force uses light gates connected to a timer or computer interface, but variations such as ticker-timers or video recordings are also acceptable. Many students lose the final mark with incomplete statements like 'the time taken for the interrupter card to cut through the light beam is measured and so the velocity can be calculated'.

(d) Explain how the acceleration can be calculated from the results. (2 marks)

(e) How can the results of the experiment be used to check the validity of Newton's second law of motion for bodies of fixed mass? (2 marks)

(f) In such an experiment the track is inclined slightly to compensate for friction. Why is this necessary for the law to be demonstrated? (2 marks)

Total: 13 marks

ⓔ A C-grade student may state that 'friction will reduce the acceleration' and so only gain the second mark. It is a common complaint of examiners that the concept of resultant force is poorly understood, and that students often state that Newton's second law requires that there are no external forces, or that inclining the track removes the external forces.

Mechanics • Electric circuits   71

Questions & Answers

> **Student answer**
>
> (a) The total mass that is being accelerated must remain constant. So if a mass is added to the load to increase the force/weight, a similar mass must be removed from the glider. ✓
>
> (b) initial velocity $u = \dfrac{2.0 \times 10^{-2}\,\text{m}}{0.20\,\text{s}} = 0.10\,\text{m s}^{-1}$ ✓
>
> final velocity $v = \dfrac{6.0 \times 10^{-2}\,\text{m}}{0.20\,\text{s}} = 0.30\,\text{m s}^{-1}$ ✓
>
> time interval between u and v, Δt = four cycles = 0.80 s
>
> acceleration $a = \dfrac{v - u}{\Delta t} = \dfrac{0.20\,\text{m s}^{-1}}{0.8\,\text{s}} = 0.25\,\text{m s}^{-2}$ ✓
>
> (c) There are several possible methods. The four most likely are:
>
> | Ticker tape | Light gate/sensor | Motion sensor | Video/strobe |
> | --- | --- | --- | --- |
> | Ticker timer (tape and oscillator) | Timer/datalogger/PC | Datalogger/PC | Metre rule/markings on the track |

ⓔ A labelled diagram can gain both marks here. ✓✓ Description of the distance measured and any corresponding time or any mention of $v = x/t$. ✓

> (d) Mention that two velocities need to be taken (or zero velocity at the start). ✓
>
> Use of $v = u + at$ or $v^2 = u^2 + 2as$ to determine a. ✓
>
> (e) Divide force by acceleration for all pairs of measurements, or plot a graph of force against acceleration. ✓
>
> Ratio F/a is constant (and equal to the mass) or the graph is a straight line through the origin. ✓
>
> (f) Newton's second law relates to the resultant force acting on the trolley. ✓
>
> The gravitational force down the slope is equal and opposite to the frictional force so that the resultant force is equal to the load. ✓

Question 15

A student was asked the following question: 'Describe how the energy of a parachute jumper varies from the moment he leaves the aircraft until he reaches the ground.'

As an answer, the student wrote the following: 'Initially the parachutist has potential energy, which is converted into kinetic energy as he descends. When the parachute opens he continues to fall at a constant rate so his kinetic energy increases more slowly. Just before landing, his kinetic energy is equal to the initial potential energy, and all of this is lost when he hits the ground.'

ⓔ This question shows examples of common errors or misunderstandings often found in examination answers.

(a) Discuss the student's answer, highlighting any incorrect or missing physics. (4 marks)

ⓔ The student who wrote the answer given would gain 0/4 marks. The following student's answer would gain the full 4 marks:

As the parachutist falls his gravitational potential energy decreases. The loss in gravitational potential energy is transferred to kinetic energy as he accelerates, and work is done against air resistance.

When the parachute opens, the parachutist falls at constant speed. The kinetic energy will remain the same, but gravitational potential energy will continue to be converted to other forms as work is done against resistive forces as the jumper loses height.

On landing, the remaining kinetic energy is transferred to the surroundings, predominately as internal energy (an increase in temperature).

(b) The parachutist falls from a height of 1000 m to 980 m before the parachute opens. Calculate his velocity at the instant that the parachute opens, assuming that at the low speeds during the early part of the fall air resistance is small and can be ignored. (2 marks)

Total: 6 marks

Student answer

(a) There is no mention of 'gravitational' potential energy. ✓

When moving at a constant speed, the kinetic energy will remain the same. ✓

There is no mention that some of the gravitational potential energy is doing work against air resistance, or being transferred to the surrounding air during the descent. ✓

The final kinetic energy will not be equal to the initial gravitational potential energy. ✓

Energy is never 'lost'. It must be converted into other forms. ✓

ⓔ Any four of these five marking points will score maximum marks.

(b) Loss in GPE = gain in KE $mg\Delta h = \frac{1}{2}mv^2$ ✓

$v = \sqrt{2g\Delta h} = \sqrt{2 \times 9.8\,\text{m s}^{-2} \times 20\,\text{m}} = 20\,\text{m s}^{-1}$ ✓

Questions & Answers

Question 16

A car of mass 1500 kg tows a trailer of mass 1000 kg along a level road. The driving force on the car is 2000 N and the total resistive forces on the car and trailer are 500 N and 300 N respectively.

ⓔ This question examines the application of Newton's second law of motion relating to a body of fixed mass.

(a) Calculate the resultant force acting on the combination. (1 mark)

(b) Determine the acceleration of the car and trailer. (3 mark)

ⓔ Many students lose marks on Newton's second law questions by failing to use the resultant force, and just including the driving force in the equation $F = ma$. You should also be aware that this force acts on the total mass of the combination.

(c) How long will it take the car and trailer to accelerate from $10\,\text{m s}^{-1}$ to $20\,\text{m s}^{-1}$ if the forces remain the same during this period? (2 marks)

(d) Draw a diagram of the trailer showing the forces acting upon it in the horizontal direction. (2 marks)

(e) Calculate the force the car exerts on the trailer through the coupling. (3 marks)

Total: 11 marks

ⓔ Once again, the key is to find an expression for the resultant force acting on the trailer alone. Since the acceleration of the trailer is the same as that of the combination, the value of the resultant force can be calculated and the pull on the trailer can be found.

Student answer

(a) resultant force = 2000 N − (500 N + 300 N) = 1200 N ✓

(b) By Newton's second law, for a fixed mass, $\Sigma F = ma$ ✓

$1200\,\text{N} = (1500\,\text{kg} + 1000\,\text{kg}) \times a$ ✓

$a = 0.48\,\text{m s}^{-2}$ ✓

(c) Use $v = u + at$:

$20\,\text{m s}^{-1} = 10\,\text{m s}^{-1} + 0.48\,\text{m s}^{-2} \times t$ ✓

$t = 21\,\text{s}$ ✓

(d) Pull of car P ← [trailer] → Resistive forces F

ⓔ Pull of car P ✓, resistive forces F ✓

(e) resultant force on trailer = $1000\,\text{kg} \times 0.48\,\text{m s}^{-2}$ ✓ = 480 N ✓

pull of car on trailer = 480 N + 300 N = 780 N ✓

ⓔ These marks could be obtained from the diagram if the forces are quantified in the diagram.

Question 17

In metallic conductors an electric current is a flow of free electrons.

(a) What are free electrons? (1 mark)

(b) At normal room temperatures, the free electrons in a length of copper wire that is not connected to a power supply have an average speed of about $500\,\text{m}\,\text{s}^{-1}$, which is due to their thermal energy. Why is there no current in the wire? (1 mark)

(c) When the copper wire is part of a circuit connected to a battery, the current in the wire can be represented by the equation $I = nqvA$, where A is the area of cross-section of the wire and q is the charge carried by an electron ($1.6 \times 10^{-19}\,\text{C}$). Explain the meanings of n and v in the equation. (2 marks)

ⓔ The question requires an explanation of the terms; simply stating 'carrier concentration' and 'drift velocity' will gain no marks.

(d) Show that the value of v is about $0.5\,\text{mm}\,\text{s}^{-1}$ in a length of copper wire of cross-sectional area $0.085\,\text{mm}^2$ that is carrying a current of $0.50\,\text{A}$, given that n is $8.0 \times 10^{28}\,\text{m}^{-3}$. Comment on this value. (3 marks)

(e) The resistivities of three commonly used materials are given in Table 7. Use the equation $I = nqvA$ to explain why:
   (i) silicon has a much greater resistivity than copper
   (ii) the resistivity of copper increases with temperature while that of silicon decreases with temperature (3 marks)

Material	Resistivity/Ω m
Copper	1.7×10^{-8}
Constantan	4.9×10^{-7}
Silicon	2.4×10^{-3}

Table 7

ⓔ The question asks you to 'use the equation', so reference must be made to drift velocity in the first case and to carrier concentration in the second. Many students give good explanations without reference to the equation and thus lose the marks.

(f) A technician wishes to make a $1.0\,\Omega$ resistor. She has a reel of copper wire of cross-sectional area $0.085\,\text{mm}^2$.
   (i) What length of wire will she need to make the resistor?
   (ii) Why would it be better to use a length of constantan wire of the same diameter to make the resistor? (3 marks)

Total: 13 marks

Questions & Answers

Student answer

(a) Free electrons are electrons that do not occupy a fixed position in the atomic structure and are able to move throughout the lattice. ✓

(b) The motion of the electrons is random, so there is no net transfer of charge. ✓

(c) n is the carrier concentration — the number of charge carriers (free electrons) per unit volume, which is directly proportional to the conductivity of a material. ✓

v is the drift velocity of the carriers — the speed at which they move along the conductor when there is a current. ✓

(d) $v = \dfrac{I}{nqA} = \dfrac{0.50\,\text{A}}{(8.0 \times 10^{28}\,\text{m}^{-3}) \times (1.6 \times 10^{-19}\,\text{C}) \times (0.085 \times 10^{-6}\,\text{m}^2)} = 4.6 \times 10^{-4}\,\text{m s}^{-1}$

$= 0.46\,\text{mm s}^{-1}$ ✓✓

This velocity is very small (compared with the random thermal speeds). ✓

(e) (i) Silicon has a much lower carrier concentration than copper and so will have a greater resistivity. ✓

(ii) In copper, as the temperature rises, the atomic vibrations increase, and so the free electrons undergo more collisions with the lattice; the drift velocity is therefore lowered and the resistivity will rise. ✓ In silicon, an increase in temperature releases many more charge carriers, so n becomes much larger, increasing the flow of charge and hence reducing the resistivity. ✓

(f) (i) $R = \dfrac{\rho l}{A} \Rightarrow l = \dfrac{RA}{\rho} = \dfrac{1.00\,\Omega \times (0.085 \times 10^{-6}\,\text{m}^2)}{1.7 \times 10^{-8}\,\text{m}^{-3}} = 5.0\,\text{m}$ ✓✓

(ii) Because its resistivity is greater, a shorter length of constantan is needed to make the resistor/the resistivity of constantan is constant over a wide range of temperatures. ✓

Question 18

An electric car has a mass of 1600 kg. When the car was travelling along a straight, level road with a speed of 30 m s⁻¹, the power was switched off and the car was allowed to 'coast' in a straight line along the road. After free-wheeling for 1 kilometre its speed dropped to 20 m s⁻¹.

(a) Calculate the average acceleration of the car during this time. (1 mark)

(b) Calculate the average resistive force experienced by the car. (1 mark)

The car is fitted with a 144 V battery that powers an electric motor. When driven at a constant speed of 25 m s⁻¹ along the same stretch of road the current flowing through the motor is 80 A.

(c) Calculate the power needed for the car to maintain this speed. (1 mark)

(d) Determine the efficiency of the motor during this journey. (2 marks)

ⓔ You will need to determine the electrical power developed by the motor.

… A-level Test Paper 1

The manufacturer quotes the average energy consumption of the car as 20 kWh per 100 km.

(e) (i) Convert 1 kWh into the appropriate SI unit. (1 mark)

(ii) Show that the energy consumption of the vehicle at the time described above is about 13 kWh per 100 km. (2 marks)

(iii) Give a reason why the calculated value is less than the stated value. (1 mark)

ⓔ You should be aware that the kWh is a unit of energy.

Electric cars are able to improve their efficiency by using regenerative braking. The motor function is reversed so that it becomes a generator and recharges the battery.

ⓔ Later in the course you will study the principles of the electric motor and generator. You should, however, be familiar with the energy transformations in each device.

(f) (i) Describe the energy transformations that take place during regenerative braking. (3 marks)

(ii) Calculate the energy transferred to the battery when the motor is used to slow the car down from $25\,\text{m s}^{-1}$ to $15\,\text{m s}^{-1}$. (2 marks)

Total: 14 marks

Total for paper: 90 marks

Student answer

(a) $v^2 = u^2 + 2as \Rightarrow a = \dfrac{v^2 - u^2}{2s} = \dfrac{(20\,\text{m s}^{-1})^2 - (30\,\text{m s}^{-1})^2}{2 \times 1000\,\text{m}}$

$a = -0.25\,\text{m s}^{-2}$ ✓

ⓔ You must include the minus sign. It indicates that the velocity is decreasing.

(b) $F = ma = 1600\,\text{kg} \times -0.25\,\text{m s}^{-2} = -400\,\text{N}$ ✓

(c) $P = F \times v = 400\,\text{N} \times 25\,\text{m s}^{-1} = 10\,000\,\text{W}$ (10 kW) ✓

(d) electrical power input, $P_\text{in} = IV = 80\,\text{A} \times 144\,\text{V} = 11\,520\,\text{W}$ ✓

efficiency $= \dfrac{P_\text{out}}{P_\text{in}} \times 100\% = \dfrac{10.0\,\text{kW}}{11.5\,\text{kW}} \times 100\% = 87\%$ ✓

(e) (i) $1\,\text{kWh} = 1000\,\text{W} \times 3600\,\text{s} = 3.6 \times 10^6\,\text{J}$ ✓

(ii) time to travel 100 km $= \dfrac{\text{distance}}{\text{speed}} = \dfrac{100 \times 10^3\,\text{m}}{25\,\text{m s}^{-1}} = 4000\,\text{s} = 1.1$ hour ✓

energy per 100 km $= 11.5\,\text{kW} \times 1.1\,\text{h} = 12.8\,\text{kWh} \approx 13\,\text{kWh}$ ✓

(iii) The car uses more energy when accelerating, climbing hills or travelling along bumpy roads etc. than when moving at a constant speed along a level road. ✓

(f) (i) Kinetic energy ✓; transformed to electrical energy ✓; then to chemical energy ✓ in the battery

(ii) Kinetic energy transferred $= \tfrac{1}{2}mu^2 - \tfrac{1}{2}mv^2$

$= \tfrac{1}{2} \cdot 1600\,\text{kg}\,[(25\,\text{m s}^{-1})^2 - (15\,\text{m s}^{-1})^2]$ ✓

$= 3.2 \times 10^5\,\text{J}$ (320 kJ) ✓

Knowledge check answers

Knowledge check answers

1. Instantaneous velocity is the value at an instant in time. Average velocity is the displacement divided by a finite time (or the sum of the initial and final velocities divided by 2 if the motion is uniformly accelerated).

2. The vertical and horizontal components of the velocity can be treated independently. Both balls will fall freely with the same acceleration ($9.8\,\mathrm{m\,s^{-2}}$).

3. The gradient of a displacement–time graph represents the velocity. The gradient of a velocity–time graph represents the acceleration.

4. The area enclosed by the line and the time axis on a velocity–time graph (i.e. the area under the graph) represents the displacement of the object in that interval.

5. **a** The graph is a straight line parallel to the time axis with a value of $9.8\,\mathrm{m\,s^{-2}}$. **b** The value of acceleration is always zero for a body moving at constant velocity.

6. The scalar and vector quantities given below are all used in this guide. There are many others that will be used later in different topics. Basic scalars: mass, distance, time. Derived scalars: speed, work, energy, power. Basic vector: displacement. Derived vectors: velocity, acceleration, force.

7. The horizontal motion can be treated independently of the vertical motion. There is no component of g in the horizontal plane and so, if no resistive forces act on the body, it will continue to move with constant velocity (Newton's first law). In reality, air resistance will act as a resistive force.

8. Equilibrium occurs when the sum of the components of the forces acting on a body is zero in any plane.

9. Balance the stand on a narrow-edged pivot so that it is horizontal. The centre of gravity will be directly above the edge of the pivot in the middle of the rod at that point.

10. The mass of a body depends on the quantity of matter within it, and so it is the same at any position in space. The weight is the gravitational force acting upon the body (mg). The gravitational field strength, g, on the Moon is six times smaller than that of the Earth so the weight will be six times smaller.

11. First law: a body will remain at rest or move with uniform velocity unless acted upon by a resultant external force. Second law: for a body of fixed mass, the acceleration is directly proportional to the resultant (net) force applied to the body. Third law: if a body A exerts a force on a body B, then body B will exert an equal an opposite force on body A.

12. $p = mv = 0.050\,\mathrm{kg} \times 220\,\mathrm{m\,s^{-1}} = 11\,\mathrm{kg\,m\,s^{-1}}$

13. moment $= F \times d = 12\,\mathrm{N} \times 0.25\,\mathrm{m} = 3.0\,\mathrm{N\,m}$

14. At each extreme position the pendulum is momentarily stationary at the highest level of the swing. The kinetic energy (KE) is zero and the gravitational potential energy (GPE) is at its maximum value. As the pendulum swings towards the mid-point, GPE is transferred to KE. At the mid-point, the GPE has a minimum value and the KE is at its maximum value. After passing the mid-point KE is transferred back to GPE.

15. By the law of conservation of energy, the total energy (or work done) in a system must be constant, so it is not possible for the output energy to be greater than the input energy (you cannot create additional energy).

16. Three $15\,\Omega$ resistors in parallel are equivalent to a single $5\,\Omega$ resistor, and two $18\,\Omega$ resistors behave like a $9\,\Omega$ resistor. The total resistance is therefore $14\,\Omega$.

17. $E = VIt = 230\,\mathrm{V} \times \dfrac{230\,\mathrm{V}}{20\,\Omega} \times 300\,\mathrm{s} = 790\,\mathrm{kJ}$

18. $\rho = 1.8\,\Omega \times \dfrac{\pi(0.25 \times 10^{-3}\,\mathrm{m})^2}{2.00\,\mathrm{m}} = 1.8 \times 10^{-7}\,\Omega\,\mathrm{m}$

19. $V = 0.40\,\mathrm{m} \times \dfrac{12\,\mathrm{V}}{1.00\,\mathrm{m}} = 4.8\,\mathrm{V}$

20. The two cells in parallel behave like a single cell of emf $2.0\,\mathrm{V}$ with internal resistance of $0.25\,\Omega$. When connected in series with the third cell, the total emf will be $4.0\,\mathrm{V}$ and the internal resistance $0.75\,\Omega$.

21. Using the equation $I = nqvA$, it can be seen that, when I, q and A are constant, the drift velocity, v, is inversely proportional to the carrier concentration, n. The carrier concentration in metallic conductors is much bigger than that in thermistors/semiconductors so the drift velocity will be much smaller.

22. The resistivity of a material with a positive coefficient will increase when the temperature increases, whereas that of a material with a negative coefficient will decrease as the temperature rises.

Note: Page numbers in bold indicate key term definitions.

A
acceleration **6**
acceleration due to gravity (g) 7–8
acceleration-time graphs 11
A-level test paper 64–77
AS test paper 51–63
average velocity **6**, 13

B
batteries 43–44

C
carrier concentration 46
centre of gravity **18**
charge carriers 45
charge (Q) 34
circuits
    parallel 35–37
    series 34–35
command terms 49
conservation of linear momentum **25–26**
    Newton's third law 27–28
contact forces 16
core practicals
    acceleration of a freely falling object 8
    electrical resistivity of materials 40
    emf and internal resistance of electrical cell 45
current in conductors, factors affecting 45–46
current (I) **34**
    in conductors, factors affecting 45–46

D
delocalised electrons 45
diodes 39
direct current (DC) 34
displacement 6, 11, 13, 14
displacement–time graphs 10
distant forces 16
drift velocity (electric charge) 45–46

E
efficiency **32–33**
elastic potential energy (EPE) 31
elastic strain energy 31
electrical energy **37**
electrical power **38**
electric current **34**
    in conductors, factors affecting 45–46
electromotive force (emf) **43–45**
energy **30**
    elastic potential 31
    electrical 37
    gravitational potential 30–31
    kinetic 31–32
equations of motion 6–7
exam guidance 48–50

F
forces 16–23
    electromotive force 43–44
    Newton's laws of motion 17–20, 21–23
    turning forces 28–29
    unit of force 21
free-body force diagrams 18
free electrons 45

G
Galileo Galilei 7
gravitational field strength **21**
gravitational potential energy (GPE) **30–31**
gravity
    acceleration due to 7–8
    centre of 18

H
horizontal motion 9

I
instantaneous velocity 10
internal resistance (r) 43–44

K
kinetic energy (KE) **31–32**

L

light-dependent resistors (LDRs) 42, 47
linear momentum, conservation principle 25–26

M

mechanics
  efficiency 32–33
  energy 30–32
  forces 16–23
  momentum 23–28
  power 32
  rectilinear motion 6–12
  scalar and vector quantities 13–16
  turning forces 28–29
  work 29–30
metallic conductors 45
moment of a force 28
  principle of moments 28–29
momentum 23–24
  conservation of linear 25–26
  and Newton's laws 26–28
motion
  Newton's laws 17, 19–23, 26–28
  rectilinear 6–12

N

newton (N) 21
Newton's first law of motion 17
Newton's second law of motion 19–20, 26
Newton's third law of motion 21–23, 27–28

O

ohmic conductors 38–39
Ohm's law 38–39

P

parallel circuits 35–37
potential difference (V) 34
  terminal 44
potential dividers 41–43
potential energy 30
  elastic 31
  gravitational 30–31
potentiometers 42
power 32
  electrical 38
practicals (see core practicals)
principle of moments 28–29

R

rectilinear motion 6–12
resistance (R) 34
  internal resistance 43–44
resistivity 39–40
  variations in 46–47
resolution of vectors 15–17

S

scalar addition 14
scalar quantities 13
Scott, David 7
semiconductors 46
series circuits 34–35
'suvat' equations 6

T

temperature coefficient of resistivity 46
tension forces 16
terminal potential difference 44
thermistors 39, 42, 46
turning forces 28–29

V

vector addition 14
vector quantities 13
vector resolution 14–16
velocity
  average 6, 13
  drift (electron charge) 45–46
  instantaneous 10
velocity–time graphs 10–11
vertical motion 9
voltage (potential difference) 34

W

weight 21
work done (W) 29–30
  electrical energy 37
  and potential difference 34, 35